PROJECT MANAGEMENT GUIDE

Marc Ducros
Gabriel Fernet

2010

t Editions TECHNIP 1 rue du Bac, 75007 PARIS, FRANCE

From the same publisher

The Oil & Gas Engineering Guide
   H. Baron

Oil and Gas Exploration and Production-Reserves, Costs, Contracts
   Centre of Economics and Administration (IFP-School)

Well Production Practical Handbook
   H. Cholet

Well Completion and Servicing
   D. Perrin

Drilling Data Handbook
   G. Gabolde, J.P. Nguyen

© 2010, Editions Technip, Paris

All rights reserved. No part of this publication may be reproduced or transmitted in any form or by any means, electronic or mechanical, including photocopy, recording, or any information storage and retrieval system, without the prior written permission of the publisher.

ISBN 978-2-7108-0952-4

Editorial

Total Venezuela is proud to contribute to the edition of this *Project Management Guide*, intended as well for its teams as to be shared with those of the National Company PDVSA, its majority and privileged partner in the country.

This guide is the outcome of the world famous experience of the Total Group in the realisation of big projects in the field of Exploration and Production as well as in the field of Refining, altogether carefully attending to the respect of safety standards, quality requirements and cost and deadlines control.

Several professionals, former or current directors of huge exploration projects- on-shore or off-shore oil and gas production, natural gas liquefaction and refining, have gathered their experiences and expertise to write this reference book.

In the Venezuela one of these big projects has a particular importance, I mean the Sincor Project, carried out between 1998 and 2002 together with PDVSA and Statoil and now operated by the joint Company Petrocedeño.

A model of horizontal management structure, it combines the characteristics of exploration production for the extraction of crude oil with those of refining in its treatment for commercialisation; it has been one of the pioneer projects of the exploitation of heavy crude oil in the Orinoco belt. It also foreshadows the huge stakes concealed in the Orinoco belt and remains today a model of development in the Venezuela.

George Buresi
President Total Venezuela

Remerciements particuliers
à Jean-Claude Soulard pour ses conseils,
sa participation et ses remarques constructives.

Acknowledgments

We would like to express our sincere thanks for the strong support and useful comments from:
Mrs Elisabeth Ducros – Checking the English wording
M. Georges Buresi, President – General Manager of Total Venezuela (up to Dec. 2009)
MM. JM Hozanski & Karim Chaouch – Total Venezuela
Total Professeurs Associés: Jean-Pierre Cordier, Alain Quénelle, Jacques Gérard,
Pierre Bouchery.
Total E & P Project Division – Alain Serceau

Table of Contents

Preface III

1. Preamble 1

2. What is a Project? What is Project Management? 5
Definitions 5
The Objective: what is to be achieved? 6
The Plan for the Project 10
The Concerted/Collective Enterprise 12
A determined effort is necessary to carry out the Project 16
A Project development also has to remain a human objective 18

3. "Risks" – Always present for all and any Project 23
Occurrence of undesirable events during the performance of a Project ... 23
Definition of a Risk 24
Butterfly effect 25
Risks in hydrocarbons field development 25
Management of Risks 26

4. The initial phases of the Project – Decision to proceed 29
The objectives of the Project 29
The Feasibility Study (FS) 32
The conceptual studies 32
The front-end engineering/basic engineering 34
The decision to proceed 35

5. The Project Organisation 37
Human management and technical management 37
The Project Manager – who is he? 38
The Project team: multi disciplines – the importance of forecasting 38
Delegation of authority 39
An evolutionary structure 39
The Task Force approach 43
Communication within the Project 43

6. The contracting phase – Reviewing different types of contracts 47
Definition and characteristics of the contractual strategy 47
Contractual strategy – Main types of contracts 49
Contract definition and general principles 55
Pre-qualification of Contractors – Call for tender procedure 56
Operator and Contractor objectives (general principles) 58
Risks in contractual performance 58
Analysis of the main provisions of a development Contract 59

7. HSE objectives and human dimension 69
What does HSE mean? 69
The main objective of the Project 69

Safety specific aspects 71
One fundamental rule about Safety 71
Security aspects 73
Environmental aspects 73

8. Scheduling the Project – Preparing and follow-up 75
The various methods 77
Collecting the necessary information 78
Scheduling by Company and by contractors 78
Follow-up 80

9. Cost monitoring – the Work Breakdown Structure 83
Cost and accounting 83
Cost monitoring: why? 83
Cost monitoring fundamentals 85
Cost Forecast 85
Payment of contractors and suppliers 86
Reporting from the Project team 87
Cost monitoring day to day action 87
Change Orders 87

10. Quality assurance and Quality control – During the Project and up to commissioning 89
Quality at the design level 89
Quality and inspection during fabrication 90
Inspection during construction 91
Final checking: pre-commissioning and commissioning 92

11. The construction phase – Relationships with suppliers and contractors 93
Organizing the construction 93
Site selection 93
Administrative matters 94
Fabrication and prefabrication 94
Offshore specific scenario 95
Performance of the construction – the discipline sequence 95
At the end of the construction 96
Relationships with suppliers and contractors 96
Relationships with local authorities 98

12. Training for project and operation – When to start? 99
Training for the project team itself 99
Training for future operating personnel and maintenance team 100
Organising a selection of operating personnel and building a training school 100

13. Summing-up and conclusion 103
A Project is always unique 103
What is a "good" successful Project? 104

Key definitions 107

Annexes 111

About the Authors 119

Preamble

Christopher Columbus was one of the first modern Project Managers, but probably not the best organised. Churchill once said (at least it is reported): "In fact, Christopher Columbus did not know where he was, nor where he was going, how long his voyage would last, or how much it would cost, and moreover, somebody else was paying the bill".

In so saying, Churchill set out the **standard rules of Project Management which** aim **to know at any moment:**

- **Where you are.**
- **Where you want to go.**
- **How long it will take.**
- **How much it will cost**.

We have no difficulty in comparing the realization of a Project with an expedition towards the unknown. Fortunately, since 1492, Project Management has made some progress. Men have been sent to the moon, and they have even succeeded in returning.

However, in spite of this progress, it happens that even today Projects see their initial objective substantially modified during their realization. This often happens in a somewhat disorganised way and is usually associated with an uncertain delay in the completion date and an uncontrolled increase in the budget.

The object of this book is to provide some understanding and guidelines on the way medium-sized and large Projects in an international environment can be initiated and managed.

Its content is based on the collective experience of the authors obtained during the last forty years, mainly with an international oil and gas Company, the rest being spent with engineering and construction contractors.

It has to be noted that this forty-year period has seen the progressive and massive introduction of computers and communications equipment in all aspects of the work.

Concern about safety and environmental protection matters has also become a priority.

This has caused a substantial evolution of the methods of management of international Projects.

It is believed, rightly or wrongly, that the experience of the authors in the development of oil and gas fields (onshore and offshore) during this time may also provide useful inspiration for the management of large projects in other domains. It is certainly the case for the nuclear industry, the chemical industry and most process industries.

We will conclude this Preamble with three remarks:

- The realization of a Project is a "one off exercise", which means there is no reference precisely defining what has to be done and how to do it, nor assessing by comparison the performance of the Project Team.
- No matter what some may say, Project Management is not a science yet: maybe is it only an art.
- A Project is always unique.

Without ignoring the interest of feedback information and benchmarking, we believe that the content of this book will be useful to inspire ideas which will have to be adapted with a critical mind to suit a specific case; it is certainly not to be literally copied and applied to the letter with the aim of providing an "automatic" solution. Even if there are analogical or similar Projects, there are *no identical Projects* as techniques evolve and Project participants, economical, geographical, political surroundings change. In the oil and gas industry, there are no identical reservoirs; even when looking at traditional building construction, we can always identify differences and in a project composed of several wellhead platforms, even if these platforms are based on the same concept, there will be some differences.

Perhaps some young engineers wishing to be Project Managers one day would like to find here the answer to the very simple question: "How shall we become Project Managers?". We face the question from time to time at the end of our courses on Project Management. The answer is not easy. It is certainly more difficult for ladies than for men: first for statistical reasons as there are more male engineers than female engineers, but also for cultural reasons. In many countries it is more difficult to be a female leader.

This being said, it is not easy to be a Project Manager for a man either. A Project Manager has to demonstrate that he has the necessary qualifications and he must be lucky enough to be available at the right time in the organisation. Finally, he must be perceived by his management as a hard worker, able to manage a team and to negotiate efficiently to protect the interests of the Company.

The installation of a new plant is not a small thing and has to be taken seriously. Indeed, it will affect the physical and economic environment not only of the people involved in its realization and after the operation, but also of the people living nearby (Fig. 1.1).

Fig. 1.1 – Covering of a gas pipeline trench passing in the vicinity of some houses: not a universally acceptable solution. (Photo : MD)

What is a Project? What is Project Management?

◎ Definitions

In Wikipedia we found the following definition for **Project**: "**Planned endeavour (determined effort towards a specific goal) accomplished in several stages.**" And: "**In business, a Project is a concerted enterprise carefully planned to achieve a particular aim.**"

It is also mentioned that: "**Project Management**" is the discipline of planning, organizing and managing resources to bring about the successful completion of specific project goals and objectives. "To Manage" means to control things in such a way they function as intended.

We can see that **Project** and **Project Management** are interdependent and have to be examined in the same process.

The definition of **Project** refers to several concepts:

- **An Objective.**
- **A Plan.**
- **A collective Enterprise.**
- **A determined Effort.**

Let us examine each concept behind this definition.

◎ The Objective: what is to be achieved?

Every Project, by nature, has an objective.

In the oil and gas business, a typical objective is the development and the exploitation of a hydrocarbon field in order to produce the oil (or gas) reserves. The sale of this oil (or gas) will allow the financing of the Project.

The word **Project** is therefore very much (we would say essentially) associated with the realization of the industrial installations (we will call them "**the Plant**"), which will allow the production and transportation of the oil (or gas, or any other product), and which are the physical outcome identified as the objective of the Project.

By virtue of this association, the word **Project** has often been used in the past in a limited way referring only to the design, construction and start up "phases" of the **Plant**.

However, nowadays, the fact that the design of the **Plant** is very much influenced by the way it is operated and maintained obliges us to consider a "global approach" encompassing all the "phases" of the **Project**.

The word **Project** may therefore also be used to cover all the phases of the evolution of the Plant from its initial concept to its dismantling, through feasibility study, conceptual and basic design, detail engineering, construction, commissioning and start up, as well as its operation and maintenance.

ᗡ What is a Plant? What is the Plant required to do?

The starting point of the Plant definition is the "Statement of Requirements". These requirements will constitute the terms of reference to check the compliance of the Plant as realized with the initial objective.

For example, an upstream oil/gas Plant is first required to be:
- Capable of processing the crude (or the gas) produced from the oil wells, in quality and quantity in accordance with the relevant specifications, as well as handling all the by-products of this processing.
- Safe to operate.
- Environmentally safe.

Please note that we could word for word replace "upstream oil" by "water production" or "process Plant to produce chemical products". A thermal power plant follows the same requirements.

The Plant also has to be economic to operate and maintain, which assumes:

- **Equipment reliability sufficient to ensure minimum Plant availability**, i.e. the proportion of time during which the Plant is capable of working properly.

   The objective of the Operator (the company operating the Plant) is usually 100% availability outside the planned maintenance periods.

- **The efficiency of the personnel operating the Plant.**

   The Operator tries to optimize this efficiency while being respectful of the safety of the Plant. This has resulted in the development of automation, making the Plant more complex.

   Today, we can often see parts of the Plant fully operated by remote control and running without permanent operators. Today's technology, for example the replacement of pneumatic control by electronic control, offers immediate action bringing better efficiency and safety.

- **The efficiency of Maintenance.**

   The Operator is usually very much concerned by the Plant shut down duration required for routine and general maintenance operations and the corresponding costs of spare parts and specialized personnel.

   Most process plants are operated 350 days a year, and large plants with several production trains running in parallel will be able to produce for two or three years without interruption.

- **Maximum efficiency for consumable consumption.**

   The Operator looks for processes and equipment with a lower consumable consumption (energy, gas, chemicals). As an example thermal power plants have gained nearly 10% on the scale of efficiency (i.e. from 30% to 40%). In the oil and gas industry (as well as for tankers) the replacement of steam turbines by electrical motors, or diesel engines or gas turbines brings a significant gain in energy.

In the Projects performed within the company we worked with, all these requirements were listed and detailed in an official document called the "Statement of Requirements".

> *This document had to be successively approved by the appropriate levels of management.*
>
> *The Statement of Requirements was utilized during the life of the Project to check the conformity of the design with the initial intent and to assess reasons for and importance of any modifications.*

In fact, the Plant in an oil/gas field is at the same time:
- For the Process Engineer, the support of a number of process systems realizing the treatment of hydrocarbons and of their effluents at the required specifications.

  We can define a functional system as a group of equipment and material designed and connected together in order to achieve a role in the final Plant.

- For the Plant Operation Manager, the combination of functional process and utility systems.

  A major Plant may contain fifty to eighty main systems split into a few hundreds subsystems.

  The difference between a functional process and a utility system lies in the final role of the system: if the system is to treat one of the final products of the Plant, it is a process system. If it is just to help the treatment of one of the final products, it is a utility system. For example a gas dehydration system is a process system while a water treatment is a utility system.

- For the Plant Construction Manager, the proper assembly of several hundreds thousands components (over one million for large Plants).

  A gas turbine alone may comprise ten thousands pieces. It was said that the Saturn rocket (the one which sent the astronauts to the moon) had around one million pieces.

- For the Project Manager, the result of the work performed by all the parties involved in the realization of the Plant, i.e. including the engineering contractors, materials and equipment suppliers and contractors.

↷ The complexity of the Plant

In short, we can define an oil/gas treatment Plant (or a Power Plant or a Chemical Plant) as the result of the proper assembly of several hundreds

thousands specially designed pieces properly fabricated from selected materials, into functional process and utility systems, involving the contribution of hundreds of engineering and construction contractors and materials suppliers, for the purpose of achieving the processing of the crude/ gas and its effluents at the required specifications.

The research for improved process and equipment efficiency and availability has led to a substantially increased complexity, particularly with the computerization of control systems.

In addition, specific metallurgies for individual processes have often been developed, increasing this complexity.

Fig. 2.1 – A large Power Plant is certainly not as complex as a petrochemical Plant, however, it necessitates more than 12 different technical disciplines. (Photo : MD)

Moreover, the search for "zero default" applied to the conformity and quality of all these pieces and their assemblies in accordance with the specification gives an idea of the size of the task.

Some have tried to simplify this increased complexity and sophistication.

I remember one attempt to come back to some simplicity. In the eighties, one international oil company with which we had a strong relationship reacted to the fall of the oil barrel under 10 US $ by an endeavour to simplify the Plant design. They used the acronym "KISS" for "Keep it simple" aiming at developing minimum facilities for less money and time.

This idea also inspired our management and I was requested to perform the development of two marginal fields on this basis. It worked; but to my knowledge the experience was not renewed. (JCS)

◎ The Plan for the Project

Once we have an idea of the initial Plant definition and how it will evolve through the life of the Project, we can deal with the next subject: How will the Project be realized? This can be subdivided into two questions:

- How will the Plant be built?
- How will the Plant be operated?

A first approach to the Plan leads us to envisage three phases (Fig. 2.2):

- The **Project Initiation phase** during which all the information necessary to take the decision to launch the Project will be collected, analysed and presented to the proper hierarchical level.

  This phase includes:

  - Preparation of the definition of the Plant up to a level sufficient to demonstrate that the requirements mentioned in the Statement of Requirements are satisfied.
  - The conditions of its realization will be examined and assessed through the preparation of the **Project Execution Plan.**

  The **Project Execution Plan** essentially includes the **Contractual Plan** for the realization of the Plant, the corresponding **Time Schedule** and **Budget** and the **Plant Operating Policy** giving guidelines on how the Plant will be operated and maintained.

- The **Project Development phase,** during which the Plant is designed, built, commissioned and started up.

  This stage essentially includes the construction of the Plant in accordance with the implementation of the **Contractual Plan** and in parallel, as necessary, the training of the future operating personnel of the Plant.

- The **Plant Operation phase** during which the Operator looks for:

  - Maximizing production;
  - Ensuring proper maintenance;

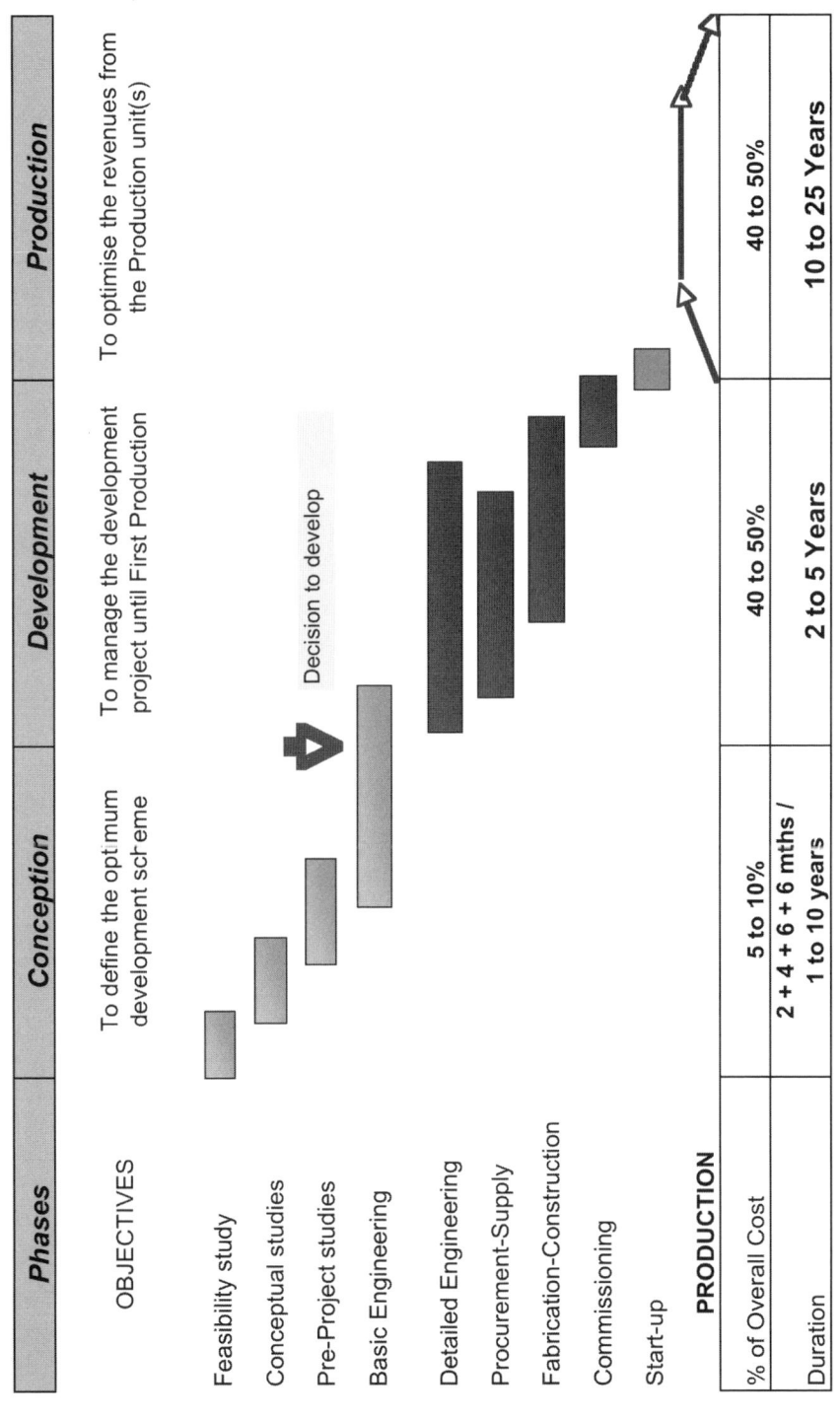

Fig. 2.2 – The Project cycle.

- Continuous adaptation of the Plant to the life of the production (i.e. oilfield or gas field or refinery) – it may include additional facilities, adapting some technology (in particular the control system), extending the Plant to match the demand for production and also the change of feedstock.
- The development of an oil/gas field is usually performed in a progressive way through successive stages. This allows the taking into account of the information obtained on the reservoir during the Production phase and therefore a better knowledge of the reservoir.

◎ The Concerted/Collective Enterprise

- **The Project is realized thanks to the contribution of a large number of different entities.** The realization of the development of an oil/gas field (and it is the same for any large project) involves a large number of actors. They come from different companies specialized in multiple domains; they are engineering contractors, suppliers of all kinds of equipment and materials, construction and services contractors of any speciality.

  They usually number several hundreds – thousands for big Projects. These companies have their head offices in different countries all over the world and may do the work in a different place from where they have their head offices. They usually have different mother tongues and cultures. Such companies may also pursue their own agenda which may not exactly fit with the Project main Objective. They employ personnel who have different backgrounds and experience in all kinds of disciplines.

This is true even for small projects.

A few years ago, for the development of a small oilfield offshore in Egypt, it happened that:

- *The management was performed by a few Frenchmen from the Paris and Cairo offices of the oil company.*
- *The basic design was performed by a French engineering company.*

> - *A US contractor realized the offshore part. It performed the engineering studies in its Singapore offices, purchased the equipment and materials in the Far East, fabricated the platform part in Egypt through an Egyptian contractor and the rest in the Arab Emirates, and mobilized marine equipment from the Arabian Gulf to perform the offshore works.*
> - *Another US contractor realized the onshore part. It performed the engineering studies in its London offices, purchased the equipment and materials in the USA and Europe and the pipes for pipelines in Mexico. It was associated with an Egyptian contractor for the construction of the facilities.*

- **The contribution of all the actors to the realization of the Project has to be organised through a Contractual Plan.** All these actors intervene to perform the different aspects of the Project realization:
    - Engineering activities.
    - Supply of equipment and materials.
    - Construction.
    - Commissioning activities.
    - Other services.

The relationship between the actors has to be organised through contracts. The contracts define the scope and responsibilities of each party, the corresponding specifications, the time schedule and the compensation. These contracts can be of any kind, e.g. contracts for provision of services or contracts for realisation of specific items of the Plant

For example we may prefer to contract:

- The Basic Engineering with an "at cost contract" plus a maximum target price, because we have difficulties to define the scope of the Basic Engineering.
- Directly by the Company for some critical items with long delivery and place the remaining orders through a general contractor.
- On a lump sum contract for the main construction including detailed engineering, supply and procurement (with the exception of the above critical items).
- Using daily rates for another part of the construction-installation work.

- Using daily rates for assistance for commissioning.
- On a lump sum basis for the temporary facilities.

The general organization of the contracting is made through the Contractual Plan (Fig. 2.3) which is an essential part of the Project Execution Plan.

We will see that the Contractual Plan has to be consistent with the characteristics of the Plant, the resources of the Operator and the availability of competent contractors and suppliers.

Some say that the strategy as expressed by the Contractual Plan is worth 80% of the success of the Project realization when its execution represents only 20%. Is it true?

It is sure that the quality of the Contractual Plan is essential for the good realization of the Project.

- **The Research and selection of experienced and competent contractors and suppliers is a necessary step; the amount of local content has to be taken into account.**

The good realization of the Project is very much dependent on the selection of capable and motivated contractors and suppliers. This requires that a survey of the capability of all potential contractors and suppliers be made in order to pre-qualify them for the performance of the Project. This survey reviews their organizations, resources and references on similar projects, as well as the workload anticipated for the next period. The most capable will be authorized to bid for the corresponding contracts. The political context usually imposes the taking into account of local content in the performance of the work, i.e. that the share of contractors from the country where the Plant site is located be maximized in the attribution of contracts. It may happen that these contractors don't always have the level of expertise and competence required. It is then necessary to support these contractors in order to enable them to do their work. This is usually not without consequences on the Project Time Schedule and Budget.

- **The negotiation process in the preparation and award of contracts and in their execution is always complex and requires a lot of attention.**

The placement and the implementation of contracts always requires a great deal of negotiation.

A successful negotiation always relies on the good communication between the parties. *We believe that satisfactory communication between people of different mother tongues and cultures is always somewhat miraculous.*

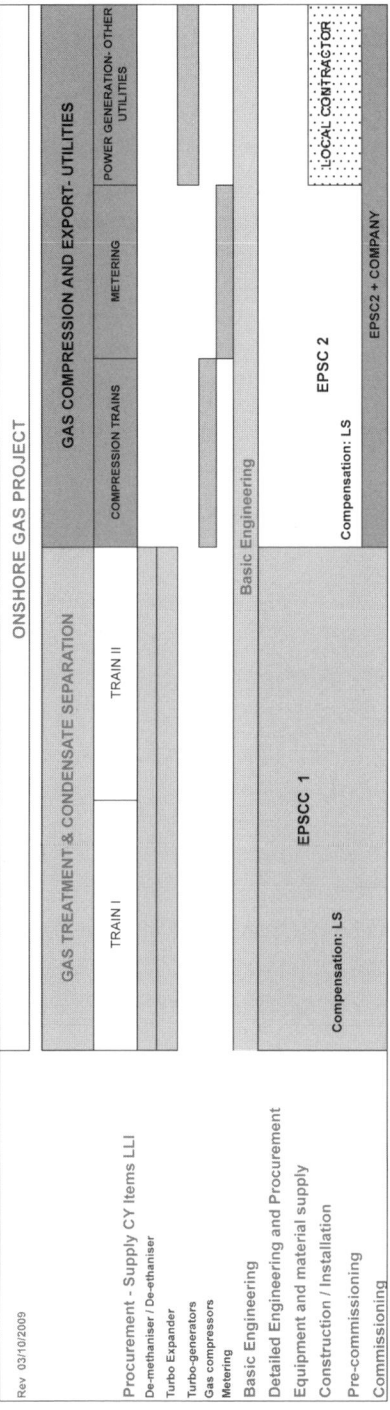

Fig. 2.3 – Contractual strategy example.

The negotiation process usually comprises verbal exchanges in meetings with the record taken in writing in order to make sure that the parties before agreeing have the same understanding of what they are going to agree.

It is fortunate that the English language is more or less well spoken all around the world, which allows basic communication.

Anecdotally on the subject: we had a kick off meeting with a fabrication subcontractor in Iran. Our representative introduced everybody from our side, including the writer of the minutes of meeting.

The representative of the subcontractor immediately reacted saying that: "If you write what we say, we are not going to say the same thing". (JCS)

Important: the magnitude of the information contained in a single EPC Contract should not be forgotten: fifteen to thirty volumes of five hundred pages each. This gives an idea of the size of the objective to be reached. This cannot be achieved without careful preparation and implementation.

◎ A determined effort is necessary to carry out the Project

- **Quasi irreversibility of the Project execution when launched.**

The decision to launch the Project is made at the end of the Initiation phase, after a lot of investigation and studies made to assess the conditions of its realization.

This means that once taken, the decision will not be overturned.

In my thirty-seven years of professional life, I have only seen two cases when the execution of the Project was suspended for a few weeks, but restarted after a few modifications. (JCS)

This means that the Project Team has its back to the wall.

- **The realization of the Plant is always a challenge due to its complexity.**

The realization of the Plant is always a big challenge. This is due to its complexity associated with:
- the uncertain evolution of the Plant definition;
- the involvement of a large number of different parties with different background and experience;
- the constraints represented by the safety and environmental precautions, the time schedule, the budget and others;

- the undesirable events which will inevitably happen;
- the risks which have to be managed.

- **The Project realization is a onetime exercise with its own terms of reference.**

The realization of a Project is a one off exercise, meaning that it is always a "première".

For hydrocarbons field development, this is related to:
- The specificity of the field characteristics that control the Plant requirements and the Plant site conditions.
- The unique circumstances of the Plant realization, including the selection of its actors and the timing of such realization.

A Project has to define its own particular terms of reference, even if this can be mitigated by the taking into account of critical feedback information and experience drawn from previous projects.

- **Time clock pressure and the Project's own pace.**

The decision to launch the Project is always made with a reference time schedule and completion date. This time schedule never incorporates float that would allow some time for additional thinking. This leads to the fact that as soon as the decision is taken, time starts running. The progress of the realization of the Project is then regularly assessed against the reference schedule. Pressure is on the Project Manager and his team to act and make as much advancement as possible.

- **Dedicated task force to manage the Project.**

Operators usually appoint a Project Manager to be in charge of all aspects of the realization of the Project. This Project Manager then designates the personnel assigned to the Project. This assignment is usually full time meaning that these personnel have to be fully committed to their work, making them a task force dedicated to the Project realization.

- **Challenge management.**

Without trying to be reminiscent of Christopher Columbus, the management of the Project has to try and know at any moment:
- What is the status of the work in quantity and quality.
- What work has already been realized.
- How much has already been spent.
- How the rest will be performed.

- How long this remaining work will take and if it is compatible with the reference completion date.
- How much the realization of this remaining work will cost and if it is compatible with the reference budget.

Therefore, it is easy to understand that the management of a Project always requires a forceful approach.

As the "where we are" rarely corresponds to the "where we should be", the Project Manager and his team always have to demonstrate their capability to imagine and implement corrective measures.

This is made through a mix of:
- Problem solving approach.
- Organization.
- Negotiation.
- Continuous check of the outcome.

◎ A Project development also has to remain a human objective

⌒ Main human objectives of a Project

What are the main human objectives of a Project today? This could be the very first question to be asked to top management and also to anyone in charge of a Project development. "To produce revenues for the Company" – A very good answer indeed… 20 years ago.

Today it is totally different: producing energy or new products must be achieved:

- By implementing Projects strictly following safety and environmental rules.
- By organizing specific training during the Projects (safety, operations, maintenance, technology transfer…), in particular on behalf of the country where the Plant is built and will be operated.
- By ensuring that future teams, in charge of the construction and the operation, can work efficiently and safely, not only for a short period of time but also for the overall life of the Project.
- By having a consistently ethical approach, specifically when selecting contractors and suppliers. The chairman of the Company often publishes a

chart. This chart usually underlines that suppliers and contractors will not be selected only for the price they propose but also for their honesty, quality.
- To complete the Project on time and within the Budget and to achieve a high quality level in order to obtain from the Plant the expected production flow.

Let us take a first example.

> *Example 1 – A new oil storage terminal on an island.*
>
> Back in 1977, a very large project was planned on an island in the Persian-Arab gulf. The first team to go there was a technical team to look at temporary facilities and soil condition. A soil survey was performed on one side of the island and a Korean company started to build their temporary accommodation on the other side. The soil survey required some borings and even if the local impact was limited, it was necessary to use engines to move personnel and equipment and to perform the job. Unfortunately thousands of superb cormorants were nesting on that island.
>
> All the engineers and managers were concerned about the birds, but the decision was in the hands of the Owner. The disturbance during the survey campaign was not enough to chase the birds away, but it was obvious that, as it eventually happened, all the birds would try to find another home. The end result was a very large crude terminal, and there was no room for all the birds.
>
> What could the situation be today (or even starting from 1995)? Maybe the birds would also be finally chased away from the island or oriented towards a specific reserve, but before doing work at site, the priority would be to perform an EBS (Environmental Base Study) often part of an EIA (Environmental Impact Assessment). These studies, performed by dedicated teams, most of the time by an external independent contractor, will carefully review all the current site conditions, the fauna, the flora, the existing level of pollution, human activities, if any, and will assess on a long term basis the impact of the new facilities. Mitigating actions will be evaluated and implemented.

⌒ Human factors impact

As we will see in this guide, the human factor is not only fundamental for the Project organization, but also for the selection and then the training of the future operators in charge of running the Plant.

The Project and the project management are based on human behaviour, human knowledge, human organization. The human factor is everywhere and it is unrealistic to imagine realising a Project without having such a basic fact in mind.

Anyone could argue: "yes, but management is also decision" and "how can you decide on a timely process, if you first need to gather the opinions of all the people involved?" The answer is easy: what will happen if you decide something either not acceptable to or not understood by the Project team? What will happen if the decision is taken on some vague or wrong information?

All actors in the Project will work more efficiently when a decision is perceived as a shared decision. We will see in this guide how to communicate efficiently within the Project team.

We can here make reference to Michel Soriano's recent book ("Vous êtes chef ?, Ce n'est pas si grave", Maxima Editeur): *"The manager must know how to delegate tasks and decisions right down to the lowest level. This will add value to the work of all collaborators who will be more satisfied with their work and will see their competence recognized."*

The important words here are "adding value". By working as a team and not as individuals we are adding value to the work, and to the Project.

But very often, as a manager, we hesitate: how shall I delegate? We may be facing the risk that the work is not performed in the same way as we would have done. Correct, but maybe the manager will learn something from delegating: a new method, better use of the computer, new presentation for the documents. The manager is also able to review the work done and to check the quality.

⌒ Using the same language

Defining certain key words to be used by all parties is a fruitful step. A common understanding may avoid disputes. In the final appendix, we have listed a few definitions that could be used for most Project developments. The experience demonstrates that, very often, standard wording is perceived differently. That is why any contract contains a specific chapter called: "Definitions".

- It is fundamental to keep the same language between the various parties working for the Project: owner, operator, contractors, suppliers, and administration…
- Good communication is part of the management role. Many disputes and conflicts can be avoided by exchanging honest information: being factual is a must. Do not hesitate to organize a meeting to clarify discrepancies.

- Using email, SMS, fax and the like, means misunderstanding is not uncommon: using the same key words helps to avoid unnecessary problems. All telephone exchanges need to be confirmed immediately by a formal written answer.
- Co-ordination documents and procedures must be developed and agreed as soon as possible between the various parties. Co-ordination procedures must be strictly applied.

Example 2 – Operator or Manager?

When it's the first time you work in a specific country and you start to deal with a National Company either as a Client or a Partner, you will certainly be facing the difficulty to be recognized as "equivalent" to the National Company. You may be called "Contractor", while you perceive your role as an Operator, or a Manager. Of course "Contractor" doesn't have a negative meaning in itself, but what do you call your "contractors" then, if you are a client: these companies doing the work at Site, supplying material and equipment…? It maybe difficult to justify in front of the National Company that you are not a contractor, as such.

Example 3 – Not speaking the same language.

Before working for an oil and gas Company, we worked in Morocco for a German Company involved in the development plan for a large chemical plant. The Client was Moroccan and the language of the contract was the French language. This was one of the reasons for my selection. The German Company was really struggling with this contractual obligation, but very often discussions took place in English, which suited both Client and company for technical clarifications. However, during weekly progress meetings, as soon as the discussion reached a certain level of difficulty, the Client often wanted to switch to French and the German Management team had to choose either to be very blunt and reject everything, or simply to accept the Client's position: using the French language effectively was too difficult for them.

A common mistake is, of course, the differences between US English and UK English: "billion" for example maybe perceived differently – like "milliard" in UK but like millions of millions in US.

> On Wikipedia we can find a list of several dozens ambiguous words. At least some are commonly used in the project management scope: "major", "lounge", "lot", "leader"…
>
> But the trouble is even worse when people from different nationalities use the English language while it is not their mother tongue. Serious misunderstandings may occur.

⌒ Ethics

The "ethical" policy in a Company is nothing new, but due to the need to fight against corruption, all main companies in the world have started to set up a specific Ethics Organisation. One of its goals is to help the Project Managers to act in a professional way.

The ethical attitude of the Company is driven by a pro-active approach: information and measuring performance.

But ethics are also for everyday business. Inside the Project team, the manager must be the guarantor of honesty and ethical behaviour: no rumours, no cheating, and no hiding of bad news. It is fundamental to recognize the merit of those working in your team. Outside the Project team, an ethical attitude requires differentiation of contractors and suppliers on nothing but objective criteria: schedule, quality and price.

What can you do if a member of your team does not have an ethical attitude? Give him a chance to modify his behaviour. So you have to tell him very clearly what the problem is, face to face, and to inform the person that without a drastic return to the normal, you will have to ask him to leave the Project.

But before meeting him, try to evaluate the seriousness of the problem. The reason being that you must first think about the reasons:

- Why is the person acting like that?
- Is he the only one in your team acting in such a way?
- Have you made a mistake? Hidden information? Passed a wrong message? Disclosed some confidential information?
- Who in the hierarchy will have to confirm your decision?

What if this person starts to threaten you? It will happen very often. He has nothing to lose and may want to take the leadership upon you…

"Risks" – Always present for all and any Project

At any time the Project is facing risks, creating risks and… can be stopped by some risk suddenly coming true.

◎ Occurrence of undesirable events during the performance of a Project

In the normal context of the realisation of a Project such as for the development of an oil/gas field, a number of undesirable events will be encountered.

To imagine the contrary would be like thinking it is possible to cross Paris by car at rush hour finding all the traffic lights are green and without encountering any traffic jams.

We do not believe in Murphy's Law: "Everything which can go wrong will go wrong". However, in the performance of complex projects, we think it may be prudent to act as if it was true.

There is also the third law of Paddy: "If you believe that everything is going right, it is because you don't know what is going on."

It deserves some consideration because it correctly reflects the difficulty of having at any moment an accurate view of the exact status of the realization of the Project.

I can recall some cases illustrating typical examples of undesirable events:
- Engineering documents, equipment and materials delivered with a substantial delay.

Once in the Arab Emirates, we had to construct the decks (upper parts) for 20 wellhead platforms in a fabrication yard. The management in optimistic mood and unwilling to risk cumulative delay on the fabrication of the deck structures and the installation of the equipment inside, had let the structures be fabricated without being sure of the timely delivery of the equipment and materials to be put inside.

The corresponding engineering documents were delayed as well as the delivery of the equipment and materials.

The empty structures already fabricated jammed the yard and paralyzed its activity. We had to organize an emergency transfer to another yard where they could be stored until the engineering documents, equipment and materials were delivered. This caused a lot of expense not to mention delay... (JCS)

- An error in the design of a Plant functional system prevents its normal operation.

  On a project in Egypt, we realized that the vibration dampeners of gas compressors packages had been designed too small and therefore could not fill their role, thus preventing a gas Plant from exporting gas. (JCS)

- Equipment suppliers, and/or contractors default due to insufficient productivity or financial problems.
- Extreme meteorological conditions causing damage.
- Major accidents.
- Major contractual disputes.
- Plant start up and operating problems.

◎ Definition of a Risk

The British define a risk as the "combined effect of the probability of occurrence of an undesirable event and the magnitude of the event" (BS 4778).

Considering the meteorological context, it addresses as well an exceptional event with most damaging effect, but improbable occurrence such as a cyclone as events with a big probability but less damaging effects, such as a number of storms.

◎ Butterfly effect

The "Butterfly effect" is the occurrence somewhere of a hurricane caused by the flapping of the wings of one butterfly at the other end of the world.

At another scale, it may happen, especially in the offshore construction environment, that a complete loss is caused by a minor event.

During the seventies, it was said that one third of the offshore pipe laying operations finished with a significant insurance claim and a major loss.

◎ Risks in hydrocarbons field development

In hydrocarbons field development Projects, the risks faced by the Operator are numerous and have potentially serious consequences.

They include:
- Field reserve risk, i.e. the field may not have the anticipated reserves, (Shukeir field).
- Project Completion risk, i.e. the construction of the Plant slipping out of control, (Bekapaï phase 2).
- Plant design risk, i.e. the Plant not producing according to specification.
- Plant operating risk, the Plant being unsafe or uneconomical to operate.
- Damage to the environment.
- Project economic risk, i.e. the Project costs (Plant realization and/or operating) climbing out of control (Π^1 factor in North Sea Projects during the seventies).
- Commercial risk, i.e. the trading of the Plant products is no longer economic. (Period when price of oil was at less than 10 US $ per barrel).

The importance of these risks and the possibility of the occurrence of events involving several such risks at the same time make it necessary to specifically address management of risks.

---

1. As for certain supersonic plane projects, it happens especially in North Sea that the budget of almost all Projects was multiplied by 3 (almost Pi) or more. But the price of oil was multiplied by 9.

◎ Management of Risks

The management of risks is made through:

- Identification of the individual risk.

    The Project Management team must be aware of the risks they are going to face in order to be able to avoid, reduce or balance such risks.

- Identification of parameters increasing the risks.

    There are parameters that substantially increase or reduce the risks in the performance of the Project. For example:

    - The choice of innovative technology is obviously more uncertain than the choice of proven technology.

        This choice will normally cause additional verifications to be foreseen during the performance of the work as well as contingencies on the time schedule and the budget.

    - The reliance on competent construction contractors or equipment and material suppliers with numerous references is less risky than the recourse to contractors or suppliers with no previous experience.

        It is then necessary to support these "new" contractors and suppliers with the expertise, which will allow them to perform satisfactorily.

    - Having to perform offshore works during the winter period increases the probability of occurrence of bad weather and therefore the risk of damage to the construction work.

        In such case, it will be necessary to establish and test procedures allowing quick suspension ("stand-by") or even abandonment of the work.

- Research of the best adapted party to deal with the risk. The best party to deal with a given risk is the party that is:

    - The most competent to perform the corresponding work, i.e. which has the necessary resources in terms of expertise, personnel, equipment and financial means.

        Some say that the commissioning of the Plant can only be properly managed by the future operating team, even if this is not always the most economical solution.

    - The most motivated, i.e. the party that will suffer most from the occurrence of the related events and therefore will take all measures to avoid these events.

> *However, we have seen offshore contractors who were continuing to lay submarine pipeline in limit adverse weather conditions because they knew the Construction All Risks insurance policy taken by the Operator would cover the damage repair or loss replacement.*

- Sharing of the risks through appropriate contractual conditions.

  One key task of the Project Management Team is to organize the distribution of the tasks and therefore of the risks among the contractors and suppliers.

  This distribution also defines the responsibilities associated with the risks.

  This is made through the Contractual Plan (see Fig. 5.1).

- Insurance policy.

  The management of risks also gets addressed the set up of an insurance coverage to finance the repair of damage and the replacement of losses as soon as possible.

  It also has to include the preparation of contingency plans for any risks not insured.

To prepare the list of risks

- *Before the Project development:*
    - As the amount of money engaged is limited, so is the risk.
    - But there are some risks for personnel (exploration phase).
- *During the Project development:*
    - All major risks can be encountered: political, technical, contractors…
- *After the Project development:*
    - Financially the risks decrease year after year.
    - The risk for personnel will remain.

Major types of risks:
- Specific to the site.
- Technical risks.
- Technological risks.
- Scheduling risk.
- Environmental risks.
- Contractual risks.

- Financial risks.
- Political risks.
- ...

Risks specific to the site:
- Access, transportation.
- Mining.
- Earthquake.
- Flooding.
- Monsoon, hurricanes.
- Tsunami.
- Offshore conditions.
- Local population.

Technical risks:
- Error of design.
- Error of fabrication.
- Underground – site instability – existing facilities (cables, pipe…).
- Heavy lifts.
- Loading – unloading.
- Offshore installations: jacket, deck, pipelines: some platforms were lost during installation.
- Offshore weather…

Technological risks:
- New technology selection – (possibility to correct, or to change).
- Not yet proven technology – (impossibility to correct).
- Cost of fixing the technology.
- Technology obsolete after a few years.

The initial phases of the Project – Decision to proceed

As we have already seen, the three phases of a project are equally important (see Fig. 2.2):
- The pre-project phase: optimization phase and freezing the Project Definition.
- The development phase: the engineering, the procurement, the construction are performed.
- The Production phase: the Plant is producing revenues. Maintenance and often, new development phases will be performed to keep the production at the right level.

This chapter is more specifically addressing the initial phases of the project.

◎ The objectives of the Project

The initial phases of the Project are essential and often critical. Why?

The project maybe stopped or postponed at any time if it does not satisfy the strategy and the criteria of the Company.

The initial work performed to define the Project may impact the entire life of the Project.

Many documents related to the Plant definition are produced during the early phase of the Project:
- SOR – Statement Of Requirements – document prepared during the conceptual study – describes the technical content of the Plant in order to

achieve the production objectives. Such procedure is very useful when we work for a large Company and with partners. But it could also be used for small projects without difficulties. The document issued will remain a reference for anyone. The SOR can be updated but only by the entities that at first prepared the statement.

- PEP – Project Execution Plan – document prepared during the Pre-Project phase or during the very first weeks of the Project, and putting together: project objectives, project organization, description of the base case, particular conditions, referential to be used, Project Schedule… Again one could say: "this is a procedure for a large Company". Not really as we could follow this procedure when building a family house.

- Hazid – Hazard Identification, Hazan – Hazard Analysis. We will review that in Chapter VII, but again we can underline that the methodology is not specific to large projects. In France many constructions were built between 1960 and 1990 in areas that later on were perceived as very dangerous with for example risks of flooding. Therefore the French procedure now imposes for all transactions and new constructions to review the Natural Risks (Risques Naturels) before any contract is registered.

- HAZOP – Hazard Operation review and analysis. All Companies involved in process production very well know such procedure. However we were disappointed to face strong opposition to implement such review and analysis for a new combined cycle Power Plant. The Project Manager in charge refused to follow the procedure and once it was eventually done, was more than reluctant to enforce the recommendations.

- Procedures to implement and organize the work. The main ones will be reviewed through the following chapters: progress work monitoring, technical review, change order procedure, accounting procedure, etc.

The specific objectives of the Project are defined during the initial phases (see Fig. 2.2):

- Expected production or productions: as per design and also maximal production, sometimes also minimal production. Built-up: it means the speed to achieve the design production.

- Consumables to be used: electricity and other energies, water, chemicals, steam.

- Level of quality, expected "life" of the future Plant and operations. Is it for 15 years, 20 years, 25 years or more? The life of the Project is often directly

- Operating philosophy:
  - The Operating Philosophy is a document prepared during the Conceptual and Pre-project phases which, for a given project, establishes the Company operations principles to ensure safe, efficient and cost effective future operations,
  - The Operating Philosophy explains the way in which the Surface[1] Operations will be organised and carried out taking into account inherent and specific constraints and context from the operations policy of the Subsidiary,
  - The Operating Philosophy identifies issues impacting on the installation design and requiring particular studies or definition during the engineering phases,
  - The Operating Philosophy is a reference document at any phase of the field development in all matters dealing with operations,
  - Is part of the hand over dossier to the entity in charge of operating the Plant after the construction,
  - Is a self-supporting document involving only necessary information. No specific details in other disciplines, such as Safety and Process, shall be provided unless they have an impact on the Operating Philosophy.
- Budget: it is prepared from the cost estimate. The top management approves it. It is based on the work breakdown structure (WBS, see example below). It is important to indicate the excluded items if any: for example custom duties and taxes. We can say that all costs not under the control of the Project should be excluded from the budget.

A Budget is based on a technical definition, a schedule, and has to be approved at the right level in the Company.

But during these initial phases, it is also necessary to evaluate the cost of the work for each of these preliminary phases: the top management will certainly be interested to know how much the "conceptual phase" may

---

1. We mean here all the units of the Plant but not the wells or any facilities below the ground, dealing with the reservoir.

◎ The Feasibility Study (FS)

The Feasibility Study is the very first step in the life of a Project. It has to be done quickly at a reduced cost.

- The main objective is identified (it maybe a prospect) – There are reasonable expectations that a project will be justified.
- The purpose of the FS is to collect enough information to support the concept of a new investment, the management or/and an international organisation (partners and State).
- The information to be collected will cover several disciplines: weather forecast, geography, geotechnical, technical, legal and fiscal.
- When a market has to be identified – the demand is reviewed from local off takers and international buyers…
- Hazid and Hazan studies are performed – see Chapter 6.
- Several scenarios maybe competing: it means different technical schemes.

The cost of such study should be limited, just in the range 0.1 to 0.3% of the final expected project cost. The duration to perform such study is also limited: **a few weeks maximum.**

◎ The conceptual studies

The conceptual study allows the selection between various technical choices. It addresses:

- The list and the number of production units to be finalised – we often use the concept of (production) "trains". But production is not enough, it is also necessary to foresee: the storage of final and intermediate products, the supply of utilities (power, air, steam, fire fighting, water, chemicals), buildings – control room, electrical room, workshop, administration, guard, fire station…

- The various choices for the Plant Processes to be assessed: for example when oil or gas treatment is foreseen, several solutions can be envisaged, but it is necessary to reduce further studies to a limited choice. The necessary utilities are identified – the consumptions are calculated or estimated.

- It is also the time to evaluate whether pre-investment is envisaged or not. Pre-investing seldom is the good answer. The reasons why: technology evolution may render the initial construction obsolete within 5 to 10 years, the market may change drastically, economically, the Project must remain viable whatever happens. However, it is often good practice to evaluate the future and to keep through the design choice the possibility to easily modify the Plant.

- High new technology or conventional approach. High new technology always brings additional risks, but may also give a competitive advantage. High new technology can also be the obliged way to produce. For the oil and gas business, the deep offshore technology is today used for more than 10% of the overall hydrocarbon production. But we can find other examples of the high new technology choice: airplanes, electrical motors, high-speed trains, wind farms, telecommunication…

- Transfer or transportation of the products: it is analysed in detail. It is certainly true for the Oil & Gas industry (and more specifically for the gas), but it could also be true for any liquid or solid production. Trains, trucks, boats, pipes… or even planes, all solutions are envisaged.

- When necessary, consultants are selected for the analysis of specific problems (for example detailed market review). Consultants can be very expensive but sometimes we do not have the choice. Do we have any expertise in house for specific telecommunication discipline, for corrosion, for local legal matters?

- Site topographical and geotechnical investigation is performed and results carefully analysed. EBS (Environment Baseline Study) is performed. Having a very good knowledge of the Site will drastically reduce some of the technical risks of the project. Environment is presently a key subject.

- In general an Environmental Impact Assessment Study is launched (EIA). It is well accepted today that any human action either for services, industry or agriculture, has an impact on our environment: drinking a bottle of mineral water or going on holiday by car is not free of any impact. Building a new production plant has to be cautiously analyzed to assess the various impacts not only on flora and fauna but also on humans staying in the vicinity of the future Plant. Just think about a new landing tarmac for an airport or new berthing facilities in a harbour.
- The first Project Technical quality Review (PTR) meeting is organised. How is this technical review organised? An independent team (from the Company) including a process manager and a safety manager is set up. At first this PTR methodology was perceived as a pure technical audit, hardly acceptable for experienced people. Today it is different, it is one more tool to improve the Quality of the Work. Independent people will be able to check all important aspects of the Project and to avoid late changes or trouble when the Plant is entering production.
- It is also certainly the time to prepare the economic model. Enough informations are available to evaluate the cost and the schedule. The economic model can be in line with the real situation.

The cost of such a study is of course above the feasibility's, it can be in the range of 0.5 to 1% of the final expected project cost. The duration to perform such study may last a few months.

◎ The front-end engineering/basic engineering

The front-end engineering/basic engineering will conclude the pre-project studies. They are sometimes performed or completed during the first phase of the Project.

Several key documents are produced during this phase. Whatever the type of the Project, we can list:
- Plot Plan.
- Flow Diagrams.
- Material Balance.
- Process and Utilities Diagram (P & Id's).
- Single line diagram.

- General Lay out.
- Hazardous area.
- Utilities list and balance.
- Characteristics and dimensions of all main equipments.
- Main characteristics of all equipments.
- Some critical material take-off will also be performed: quantities of pipes for a pipeline, quantities of steel for a platform, take-off for piping and valves to be installed in the Plant.

The **referential** to be used for the performance of the Project must be defined at this stage. The referential makes reference for all documents, including specifications, procedures, methods, codes to be applied by all contractors and suppliers performing for the project.

From time to time, we can also face special scenarios where previous projects have a certain impact on the present Project.

> *Example 4 – Trying to use already ordered material and equipment.*
>
> One of the first projects in which we were involved for the Oil & Gas business was a Project in the Middle East. The local Partner asked the Project Team to use compressors, drums and columns purchased three years before for a similar Project, unfortunately cancelled. The detailed studies were performed and some modifications had to be implemented on the already purchased equipment. It was therefore necessary to send the equipment back to original factories and to proceed with the modifications, then to return the equipment to the Site. In the end one could demonstrate that the cost of all the jobs was superseding the cost of brand new equipment.

The cost of the Basic Engineering is significant. It can be in the range of 2-3% and up to 5% of the Project cost. The duration is never less than six months and can last up to 15 to 18 months.

◎ The decision to proceed

The decision to proceed can be taken on the basis of the Plant definition, on technical review, the Project schedule analysis and risks evaluation, cost estimate and the corresponding economical review. The decision to proceed is joined with the reference budget approval.

But sometimes the management interrupts the process before the pre-project phases are completed (i.e. the project is too risky or does not match the economic criteria of the Company).

Several key tasks have to be completed during the pre-project phases: site geotechnical and topographical survey, safety review, project technical review, environmental impact assessment...

It is fundamental to keep in mind that the pre-project activities are essential to achieve the Project properly. Trying to by-pass some of the initial phases is dangerous and may have dramatic consequences on the budget, the schedule and the result of the Project.

Risk evaluation and risk mitigation are two important steps of the management approval.

Once the decision to proceed is taken, a Project Manager is named and a team is selected to start the execution of the Project. It may happen that the selected Project Manager was also in charge of the pre-project studies.

Unfortunately the decision to proceed can be rejected either by the management of the Company and/or by the Partners. The process of the previous pre-project studies can be resumed to achieve better solution or the process is just terminated.

We can write a long list of very important Projects delayed by 5 to 10 years and sometimes more. This means that to start a Project, many conditions need to be satisfied and sometimes only one is enough to stop the Project, or at least to postpone the decision. Most are now realized or will be realized:

- Alwyn Project in the North Sea.
- Hydro Project in Laos.
- 1st LNG Project in Nigeria.
- Bongkot Field development in Thailand.
- First Refinery project in Vietnam.
- LNG terminals in the USA.
- Gorgon project in Australia, etc.

The Project Organisation

◎ Human management and technical management

Human management, contracting and negotiating, scheduling, cost management, technical management, logistic, safety, security and environmental management… all disciplines have interfaces and have to be approached in a professional way. We do not want to oppose any kind of management but obviously we cannot oppose "Technique" and "Human value" any longer.

Human relationships must be taken into account at all levels: any manager is in charge of a team, sometimes a very small one, but it is important for the behaviour of all managers to be governed by the same guidelines.

Let us take an example: suppose that we are working in a South-Asian country; the team is naturally composed of expatriates and local staff. What will happen if the expatriates do not evaluate the local staff in the same way? What will happen if at the end of the year the bonus or the salary increase is not distributed in the same way (does not mean the same amount)? What will happen if one manager invites his staff for lunch once every three months and another once a year or even less?

Technical management is another topic. The Project Manager is not necessarily the most technically qualified within the Project Team. His expertise has to be good but he cannot be an expert in all subjects. He has to trust other experts and he has to find the way to evaluate their performance from time to time.

◎ The Project Manager – who is he?

A Project Manager is a person selected by the hierarchy to be the leader for a Project development. He is a team leader, but he will not be doing everything. He has enough experience, but he does not know all disciplines and all the details of the Project.

Rule N° 1

The "perfect" Project Manager is not only:

- *A good technician, "engineer".*
- *A good economist.*
- *An expert with a lot of diplomas.*
- *A brilliant future Director of the Company… or even more…*

But also…

- *An energetic (not authoritarian) team leader, paying enough attention to everyone working in his team.*
- *An excellent communicator with a good degree of charisma and a good sense of humour.*
- *Someone concerned by long term value with long term vision.*
- *Someone with sound moral behaviour.*
- *A leader able to anticipate the problems and the difficulties.*

◎ The Project team: multi disciplines – the importance of forecasting

Managing the Project is not the role of the Project Manager alone, it is of course a project team's overall task.

All disciplines play an important part and all need to be governed by the same management attitude: *forecasting*. Interfaces and coordination between disciplines (cost-schedule, cost-technique, technique-safety…) will guide the day-to-day work management. Forecast and anticipation are the keys for a successful Project. No one can expect to finish the Project on time and on schedule without anticipating delay from suppliers, lack of manpower from a contractor, problems with the customs, incomplete documentation, lack of spare parts, etc.

◎ Delegation of authority

Taking decision in due time is of course the best way to perform good project management. Delegation of authority is definitively a must to allow the Project Manager and his team to react promptly and efficiently. The delegation is not only meant for expenses, but also for all approvals the Project has to deal with: approval of documents, time-sheets, minutes of meetings, technical approval of the work done, approval of invoices before payment… etc. and also management of personnel.

I remember having worked for a German Company for about eighteen months. I was named deputy to the Project Director for a quite important Project in the north of Africa. Very soon I got frustrated: the delegation of authority was almost nil and it was impossible to purchase a 1 € pencil without the approval of two directors. On the other hand a draftsman could produce a very inadequate drawing without real control but with huge financial consequences for the Project.

⌒ Typical Delegation Chart

"Parties" means here below either the top management and/or the partners and the management.

◎ An evolutionary structure

Project organisation is the most important tool of the Project. The organisation will move all along the project development. The Project will start with a relatively small team. During the engineering phase and the construction phase, the team will be reinforced to keep all the activities under control. Then the Project team will be progressively demobilised.

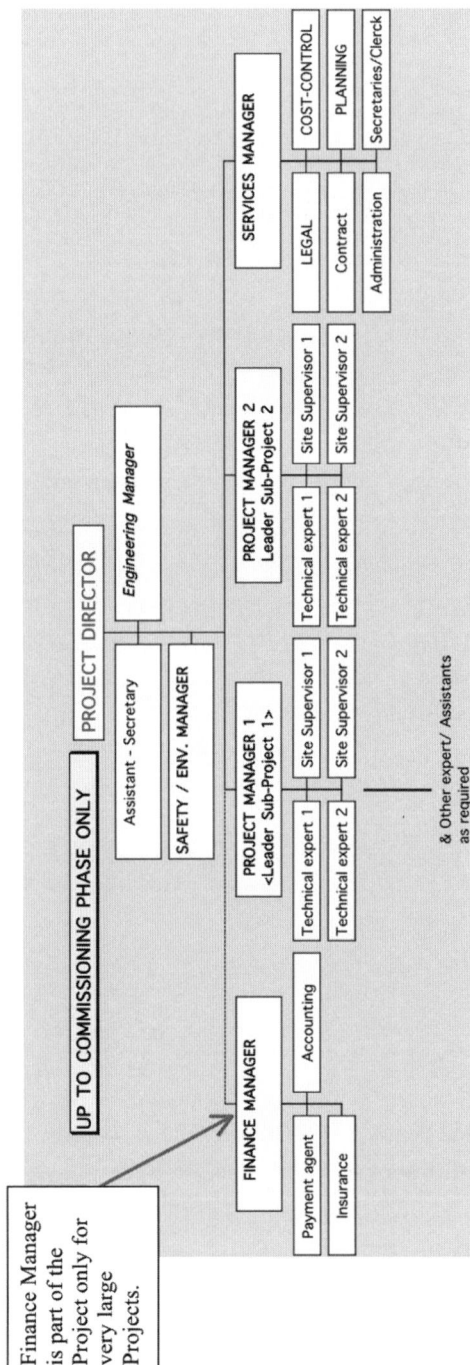

Fig. 5.1 – Project Development Organisation Chart.

Important: The Safety/Environment Manager is directly below the Project Director.

The Engineering Manager (here coordinator) is in charge of organizing all necessary studies, inside and outside the Project team, to complete the Definition of the Project.

The Finance Manager function is not always part of the Project, as the Finance Manager could be part of the Affiliate in charge.

The Project is here divided into two Sub-Projects.

The Services Manager is in charge of common services such as Legal, Cost-Control, Contract, Planning and general administration of the Project. The Legal expert is part of the Services or directly connected to the Project Director. Legal experts have taken more and more importance since the seventies. This is due to the Anglo-Saxon influence and also to the increasing complexity of many projects.

Table 1 –

	To Parties for Information	Approval by Parties	Approval Cycle (Days)	Comment
Contractual Strategy		X	10 D	May be approved by only one meeting
Bidder's list				
for estimated amount < 100 k$	X			
for estimated amount > 100 k$		X	10 D	May require a specific approval process
Prequalification (Recommendation)				
for estimated amount < 100 k$	X			
for estimated amount > 100 k$		X		Any prequalified bidder cannot be disqualified afterwards
Tender documents				
for estimated amount < 100 k$	X			
for estimated amount > 100 k$	X			Communication on demand
Bid Analysis & recommendation				
for estimated amount < 100 k$	X			
for estimated amount > 100 k$		X	15 D	
Actual Order				
for estimated amount < 100 k$	X			
for estimated amount > 100 k$	X			
Amendment				
for estimated amount < 100k$	X			
for estimated amount > 100 k$		X		
Arbitration Notice/Suite		X	20 D	Top management approval

42 *The Project Organisation*

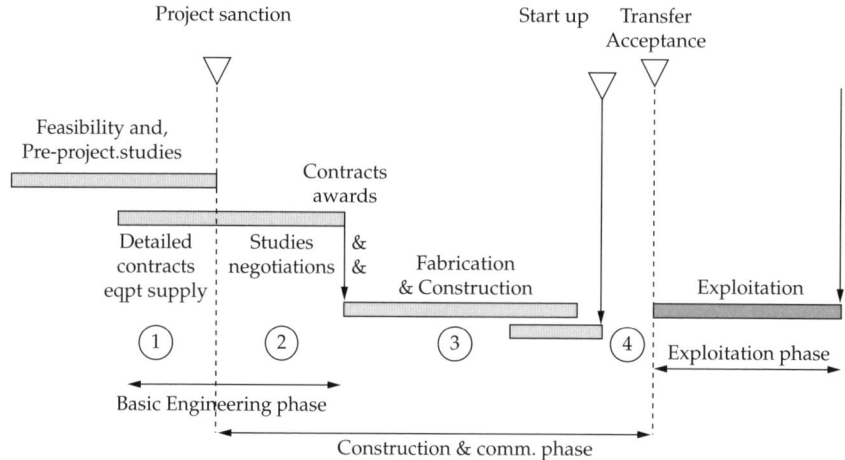

Fig. 5.2 – Evolutionary structure. Phase 1 – Preliminary studies.
Phase 2 – Basic Engineering or Front end Engineering.
Phase 3 – Fabrication and Construction.
Phase 4 – Pre-Commissioning – Commissioning.

At the start, the Project has of course to be staffed by a Project Manager but also by an Engineering Manager and a Service Manager.

The role of the Engineering Manager (it could also be named "Engineering coordinator") is to coordinate all the Project technical subjects within the Company as well as to deal with all interfaces with contractors and more specifically the engineering contractors. He is able to manage all technical specialists and to identify what is necessary and what could be "nice to have" but is not absolutely necessary for the Project. It is better to give this role to an experienced senior engineer (min. 15 years experience). As the process, the drawings and the key technical calculations are performed by professionals, the Engineer Manager must be able to identify quickly any error or mistake. He must also be able to define the work to be done by the various Parties involved in the performance of the Project.

The Service Manager or Project Control Manager is in charge of organizing and coordinating the non technical activities of the Project, such as the contracts and purchase orders preparation. He also coordinates the scheduling of the Project and the cost control of the Project. He directly acts below the Project Manager. Sometimes perceived as a secondary position, the role of the Control Manager is certainly today a key position in the Project Team.

◎ The Task Force approach

The Task Force approach is the preferred solution for large or medium sized projects.

A task force represents a unity within the project team and avoids conflict between projects sharing the same personnel. The Task Force means that a large part of the Project team is geographically at the same place and working for the same Project Manager. The traditional hierarchy for each member of the Task Force, is temporarily left aside at least for the operational aspects. Very often separate offices are arranged for the Project Team in order to avoid disturbance of any kind.

From relatively small to medium sized projects: when the Project cannot afford a permanent staff for all technical disciplines, the Task Force will be limited to the Project Manager, the Engineering Manager, the Service Manager and 3 or 4 other key people.

Team spirit is a very serious advantage of the Task Force. *A contrario* it is very difficult to create a team spirit without keeping the Project Team within a task force.

Again the structure of the Project Team is not permanently stable: the Task Force is not the same along the Project development. Mobilisation is progressive, demobilisation can be achieved within a few weeks.

The decentralized activities: some activities can easily be performed outside the Task Force. For example and with the exception of very large projects, the Finance Manager can be working at the main office far from the Project Team. For Oil & Gas Projects we do not always incorporate the manager for geophysics and reservoir inside the Project team.

◎ Communication within the Project

Communication inside the project and with third parties is analyzed and the Project Team will define how the information is distributed: documents, correspondence, data, decisions, etc.

A central filing system is also very important in order to keep the originals of all contractual documents or at least a clear copy.

We give below an example of a distribution chart for some engineering documents:

Table 2 – Typical table

Function	All contractual correspondence	Key technical documents	Specifications for main equipment	Price analysis
Project Manager	CIA	CIA	CIA	CIA
Contract Manager	CIA	CI	CI	CIA
Engineering Manager		CIA	CIA	CI
Services Manager	CI	CI	CI	CIA
Mechanical senior specialist		CI	CIA	CI

Note: "CIA" copy for all – for information and approval, "CI" copy for information.

Of course the list of documents is here limited, as well as the members of the team: it has to be developed according to the Project.

- Formal communication. Any written communication is of course the basic approach. If the dispatching is properly done, everything will be fine. But what does "formal" exactly mean, what does "communication" mean, what does "dispatching" mean? Formal refers of course to written words, but not only: the language must be appropriate and needs a semantic approach. For example the statements included in the wording need to be perfectly clear and understandable for every one. The use of conditional form is very dangerous. Communication means exchanging information between parties. Dispatching means sending communication to a certain list of addressees.
- Informal communication. It could be verbal, it could be SMS today. Any party has to be cautious when using phone or e-mails.
- Nothing like a meeting! A meeting often gives the possibility to solve big differences between Parties and to avoid conflicts. Many people can be very aggressive when writing, but their attitude face to face will be different. If meetings at the Project level cannot solve the problem, it will then be necessary to organize a meeting at the upper management level of the companies.

It is good practice to prepare minutes of meeting as soon as possible. Who writes the minutes, how the minutes are reviewed and approved is very important.

The Client/Operator should try to keep the leadership on that matter. However, very often the Project team is not staffed with many people and writing minutes may take time. In any case, the best solution is to agree on the minutes **at the end of the meeting**. The worst solution is to agree on the minutes at the next meeting, almost inevitably it creates difficulties.

- Communication towards parties outside the Project. Let us set aside the communication with the press: it is usually managed by a special team inside the Company. But communication inside the Company is also fundamental. The Project must be perceived as business bringing value to the Company. Other divisions, other departments must be convinced that the money of the Company is well invested. To achieve that, and with the assistance of the internal communication service (if any), regular information on the Project must be disclosed.

The contracting phase – Reviewing different types of contracts

◎ Definition and characteristics of the contractual strategy

The contractual strategy is prepared from the pre-Project phase. The project team has to submit the strategy to the top management.

> CONTRACTUAL STRATEGY:
> - The establishment of the Contractual Strategy is the most critical aspect during the initial phases of the project.
> - What does that mean?
> – How many contracts, Purchase Orders.
> – How to separate the units, the sub-projects. Where to place interfaces?
> – What type of contracts will be used: EPSCC, Price List, Reimbursable, Cost + Fee. Separate orders for Long Lead Items (LLI).
> – The content and definition of each contract or P.O.
> - Several key aspects have to be taken into account:
> – Legal and local rules.
> – Risk mitigation.
> – Qualification, Experience and management of potential contractors.
> – Availability of large construction equipment as needed.
> – Contractors Payment Method: progress or milestones.
> - The Work Breakdown Structure (WBS) is the basic tool (see further down).
> - Cost and schedule can be affected by the contractual strategy. Coherence is a necessity.

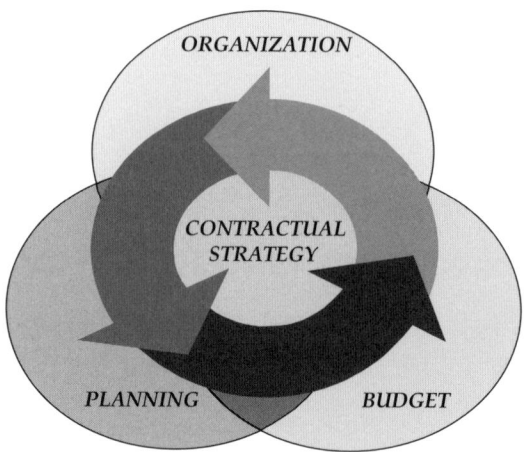

Fig. 6.1 – No activity is really independent from the others.

The make-up of Contractual strategy is an essential element of the Project Management.

In this Chapter, we will use the word "Operator" for the Company acting on behalf of a group of companies to develop the Project and build the Plant.

⌒ *Purpose of this phase*

The purpose of this phase is to define the industrial and contractual policy for the realization of the forecasted installation.

Generally, this phase takes place at an early stage of the Project: at the time (or before) of the decision to develop.

⌒ *Two Main factors to be taken into account*

In accordance with the Project environment, the main factors to be taken into account are the country or the area where the Project takes place, the political and legal rules applying to the Project, the culture and experience of the Project partners.

In accordance with the Operator human resources, the Contractual Strategy will take into account the Operator Project Management and technical resources.

Regarding the technical definition of the development, the Contractual Strategy will take into account the size and complexity of the Project to construct, the part of innovation in the development, the part of the respective tailor-made and on the shelf equipment, the interfaces with existing facilities.

The major technical choices have an influence on the Contractual Strategy and vice-versa; for instance in the Oil & Gas industry: offshore big modules versus small modules.

As for the industrial conjuncture, the following factors have to be taken into account: main potential contractors and suppliers work load, technical abilities and experience, market price level and foreseen evolution. A brushing up of the conjuncture and market is usually done through a process of pre-qualification of potential bidders.

The Project risks must also be properly appraised by the Project Management team: sensitivity to environmental constraints, field reserves evaluation, consequences of design failure, schedule slippage (consequences of production delay for Oil & Gas projects or for a Power Plant), economical consequences of cost overrun, operating hazards.

It should be noted that this phase is sensitive and could cause a form of "lobbying" from the potential contractors (see Fig. 2.3).

◎ Contractual strategy – Main types of contracts
(Figs. 6.2 and 6.3)

For development projects, in order to build the installations of the Plant, the main possible contractual options for the Operator are the following:
- Multiple separate contracts: engineering, equipment and materials, prefabrication, transportation, construction, services.
- EPSC (for engineering, procurement, supply and construction).
- EPSCC (for engineering, procurement, supply, construction and commissioning).
- Turn key contracts, for which the Contractor is in charge of everything and performs the whole realization of the Plant.

⌒ Separate contracts case

In this case, the Operator directly and individually contracts with a number of different individual suppliers and contractors to have the various steps of the realization of the Project performed.

- Basic design – an engineering contractor is in charge of performing the Project Plant Definition under the control of the Operator.
- Detailed engineering.
- Procurement (organization of purchasing activities) of equipment and materials.
- Supply (fabrication and delivery) of equipment and materials by many different suppliers.
- Construction of the Plant by one or several contractors.
- Technical assistance to the commissioning of the Plant.
- Technical assistance to the start-up of the Plant.

Each contract can be remunerated in a different way: lump sum of course, but it can also be daily rates, rates per operation, etc.

In the separate contracts case, the Operator:
- directly manages and controls the evolution of the Plant definition;
- directly places all the purchase orders and contracts;
- performs the direct control and acceptance of the equipment, materials and construction work;
- directly manages all related interfaces;
- performs the commissioning with his personnel and with the assistance of Equipment suppliers and Construction Contractors.

However we can list projects that follow such strategy as well as have the constructions of some units of the Plants performed by general contractors dealing with several disciplines. We can also remark that very often when the type of remuneration is of the cost + fee type for the main contractor, the number of subcontracts and orders to be managed is significant.

Through his direct control of the individual suppliers and contractors, the operator can:
- optimize the performance of activities going on in parallel;
- identify and promptly correct any critical delay.

The contracting phase – Reviewing different types of contracts **51**

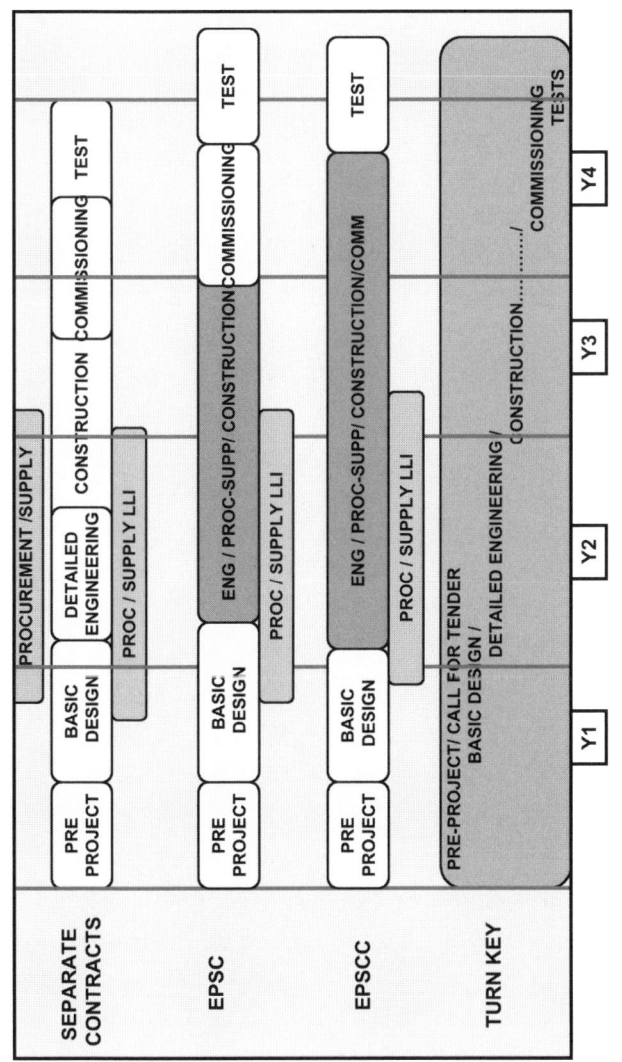

Fig. 6.2 –

LLI: Long Lead Items.

52   *The contracting phase – Reviewing different types of contracts*

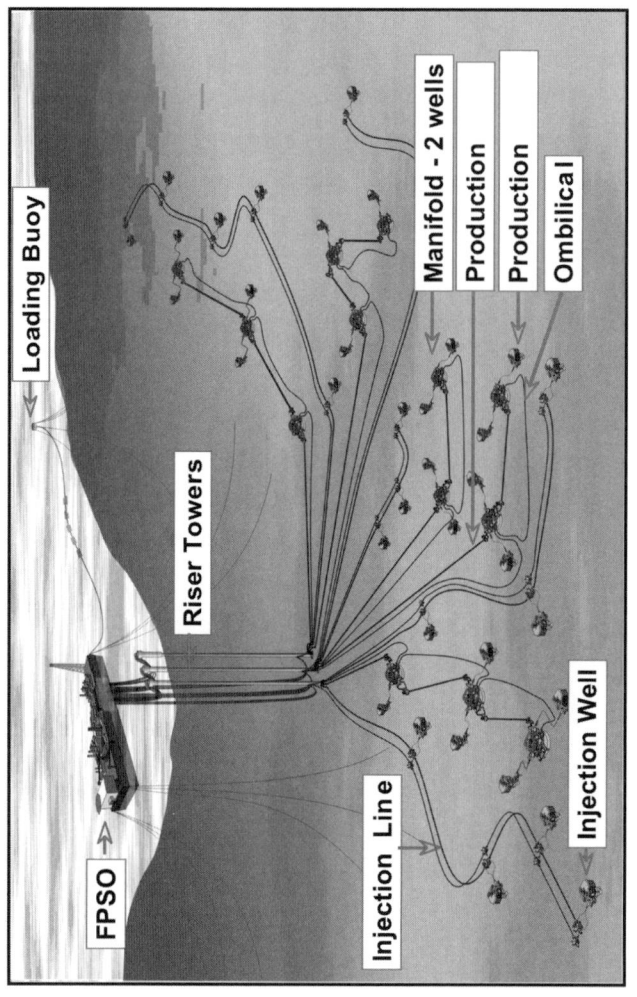

Fig. 6.3 – Girassol Project – 4 Main contracts only.

Source: Total.

Theoretically this leads to a competitive time schedule.

This solution requires a sufficient competence and capability of the Operator to manage the respective contracts and the corresponding risks and interfaces.

This approach has also been used to a certain extent for North-Sea Projects at the end of the '70s. Even in the 1990s, we have seen projects with a mix of EPC and daily rates contracts.

⌒ EPSC Contract case (for "Engineering, Procurement, Supply and Construction")

The Operator:
- performs (or gets performed) the basic design and the general engineering of the Plant, for which he remains responsible;
- may directly place orders for the Plant non standard equipment and materials, mainly those having a critical delivery (LLI – Long Lead Items);
- controls their manufacturing and shop testing before delivery to the EPSC contractor, or alternatively transfers (assigns) the orders at EPSC contract award or very soon after;
- and in parallel, places the EPSC contract.

The EPSC contractor:
- endorses the constructability of the Plant on the basis of the general engineering of the Plant;
- performs the detailed engineering of the plant and the construction engineering;
- purchases the standard equipment and non critical materials in respect of schedule;
- performs the construction up to the status of "ready for commissioning".

In this case, the Contractor is responsible for the whole construction and assembly of the Plant but he does not take responsibility for the performance of the Plant.

"Endorsement" means that the Contractor checks and takes responsibility for the validity of data and technical specifications prepared by the Operator in the tender documents as if such data and specifications were prepared by the Contractor himself.

⌒ EPSCC Contract case (for "Engineering, Procurement, Supply, Construction and Commissioning")

The Operator:

- performs (or gets performed) the basic engineering of the Plant;
- selects non standard equipment and materials, mainly those having a critical delivery;
- assigns the orders of such equipment to the EPSCC Contractor at contract award.

The EPSCC Contractor performs:

- the plant detailed engineering;
- the procurement of the orders for the remaining equipment and materials;
- the construction;
- the commissioning operations (verification that all the elements of the facilities are functioning correctly) in accordance with Company rules and methodology.

He also assists the Operator in the Plant start up and maximum performance testing for the surface/subsurface facilities.

After Contract award, the Operator's role is mainly to monitor and control the Contractor's performance of the work. The management of the project remains with the operator.

⌒ Turn Key Contract case (see Fig. 6.2)

The Turn Key Contract gives full responsibility to the Contractor from the Pre-Project phase to the start-up of the Plant. The Contractor performs the whole realization of the Plant (engineering, procurement, supply, construction, commissioning and start-up). The realization is based on an initial plant proposed by the Contractor and to be agreed by the Operator.

Such concept is not often used in the oil and gas upstream industry: usually, the Plant realization is started with, even though the information on the reservoir characteristics is incomplete, as this realization is often performed in parallel with the development wells.

For midstream (pipeline...) and downstream (Power plant, chemical and petrochemical Plant...) projects, the turn-key concept is valuable under certain conditions:

- Final objective perfectly defined.
- Possibility to precisely measure the performance of the Plant.

To sum up, the most common solutions used in the Oil & Gas upstream industry for development projects are now either the EPSC Contract case or the EPSCC Contract case.

◎ Contract definition and general principles

A "Contract" is an agreement between two or more parties by which they commit themselves to fulfil defined obligations.

We do recommend, for a new Project, to always start from the form of standard Contract and to adapt it to this new Project: starting from previous Contracts could be detrimental to the Operator and not appropriate for this new Project!

Within a development Contract:
- The Contractor commits himself to carry out activities related to the performance of the Project.
- The Operator commits himself to compensate the Contractor for the carrying out of such activities.

At any time during the performance of the Contract, the parties can mutually agree to modify the Contract conditions. In this context, it is understood that:
- Both parties must perform their respective obligations.
- Should one party fail to perform, that party is said to be in "breach" of Contract.

However, if a failure of performance is caused:
- either by an outside influence not anticipated under the Contract,
- or by the other contracting party.

the non-performing party may be excused for its non-performance.

From a practical point of view, it is commonly said "a Contract has the value that the parties accept to give it". Disregard by one party of the contractual rules may:
- Affect the performance of the work, the time schedule and the Contract price.
- Quickly lead to the modification of the contractual relationship.

The disregard of the contractual rules maybe due to the Operator or to the Contractor, with consequences that could affect the Contract completion date and price.

The contractual rules must be realistic: contractual requirements that the Operator or the Contractor cannot satisfy should be avoided.

◎ Pre-qualification of Contractors – Call for tender procedure

A development Contract is usually awarded and signed at the outcome of a "Call for tender procedure": the purpose of this procedure is to select a Contractor among several possible contractors ("tenderers").

The main objectives of this procedure organised by the Operator are:
- To ensure a fair and equitable treatment of all possible tenderers.
- To select a competent contractor among sufficient competition.

Tenderers are generally requested to submit their offer ("tender") in three parts:
- The contractual part (essentially items of non compliance with the contractual conditions proposed by the Operator).
- The technical part (essentially the qualifications to the design dossier including the "execution plan") with the means and the proposed "work time schedule".
- The commercial part with the proposed price conditions.

The Operator will hold "clarification meetings" with each tenderer. The joint review with each tenderer will allow:
- The Operator to appreciate the understanding and motivation of the tenderer in regard to the dossier.
- The tenderer to improve his understanding of the dossier.
- The Operator to improve the general quality of the dossier by taking into account the relevant comments from the tenderers.

When one of the tenderers is asking for clarification either technical or contractual, the Operator has to answer in writing and shall pass the information to all tenderers: this is the way of being fair and equitable with all bidders.

After the clarification meetings, the Operator will issue updated "Call for tender documents" taking into account the comments made by the various tenderers.

On this basis, the tenderers will be requested to submit their final tenders:
- Without any contractual or technical qualification.
- With an updated Contract execution plan complying with the required completion date.
- With a Contract price fully inclusive of all the Contract conditions.

The final tender usually includes:
- The main Contract documents already initialled by the tenderer.
- The execution plan with a time schedule compatible with the completion date requested by the Operator.
- The complementary information requested from the tenderer during the clarification meetings.
- The final fully inclusive price.

The final tender should be submitted in two parts with the commercial part separated from the other documents. Opening of final tenders and selection of the Contractor ("selected tenderer"):
- After having checked that each tenderer has accepted (through his initialling) the Contract conditions, the Operator opens the commercial proposal.
- The selection is made in principle on the lowest price basis, or at least on the most economical solution for the Operator.
- Most economical means that during the life of the Project the selected Contractor is the lowest combining Capex and Opex.

  For example a compressor can be more expensive for Capex but requesting less maintenance or being equipped with cheaper spare parts.
- If it is not the case, the Operator is usually requested to document his decision.

To accept, at the last minute, an uncontrolled price reduction from one tenderer outside the normal competitive procedure would be:
- Unfair to the other tenderers.
- Risky for the Operator (the Contractor will probably try to recover such reduction during the performance of the Contract).

This practice of "last minute reduction" is not uncommon and could raise various problems with the non-selected tenderers.

◎ Operator and Contractor objectives (general principles)

The Operator aims at ensuring that the realization of the Plant and its operating be economically sound.

The main objectives pursued simultaneously by the Operator during the Plant realization are: safety and environment protection, satisfaction of all the technical requirements (specifications), timely completion, and costs within budget.

Regarding the technical requirements, the objective of the Operator is that the Plant be "fit for purpose". "Fit for purpose" means that the Plant, when properly operated is able to satisfy all the technical requirements in respect of the operating performances and characteristics.

Obviously, the Contractor does not completely share the same objectives as the Operator:

In trying to minimize his risks, the Contractor aims at ensuring the continuity of his existence: he is essentially concerned, at first, by obtaining contracts and by making them the most profitable.

The Operator may have a budget margin at his disposal, which allows him to be more flexible than the Contractor.

In any case, it is important to develop between Operator and Contractor what we could call a win/win relationship where the interests of both the Operator and Contractor are safeguarded on an equitable basis. This is often an essential key for the success of a development project.

◎ Risks in contractual performance

The realisation of a development project is hardly ever automatic and it generates numerous problems.

Risks of problems to be faced during the performance of a development Contract are increased due to the following factors: the complexity of the scope and technical specifications, the length of the realization, large number of entities involved in such realization, usually settled in many different countries, the large number of external events, such as earthquakes, floods, wars or strikes, which can potentially affect the realization of the Project.

The "disturbing" events must be faced by the Operator and the Contractor with a "solution finding" approach. Both parties must find an appropriate solution to the consequences of the events affecting the performance of the work. The main purpose of this "solution finding" approach is to try to mitigate their effects on the Contract time schedule and the price.

This "solution finding" approach should first be made on the basis of the Contract documents. Then, if the issues are not solved, some kind of "negotiating" between the parties is required at the project level. Then, if not promptly resolved, it is usually submitted to the upper management of each party for review and resolution. If not settled at this stage, either party may decide to go to arbitration or to court.

The most common claims or disputes between Operator and Contractor are based on or linked to:

- The Plant site conditions or access of the Contractor to the Plant site not as foreseen in the Contract.
- Interferences by the Operator in the management of the Contract, for instance the disturbance of the performance of the work by the Operator's personnel or other contractors.
- The modifications of the Contract scope during the performance of the Contract.
- The failures by the Operator to timely pay the Contractor's invoices.

◎ Analysis of the main provisions of a development Contract

We are going to examine some of the main provisions of a typical development contract, being understood that there are many other provisions in a development Contract. The following provisions are given as examples only:

⌒ Contract law and arbitration

Applicable law under the Contract: The applicable law should be specified in the Contract. The choice of the law can affect the substance of the Contract, for instance the duration of warranties and guarantees. The choice of the law can even affect the Contract price if the law selected contains a lot of uncertainties: the Contractor selected could increase his price to cover such uncertainties.

The applicable law should be the same for all the contracts of the same development Project for instance, in case of litigation involving the Operator and two contractors of the same development Project. In addition, the Operator should select an applicable law with which he is familiar. The provisions of the applicable law apply even if they are not individually set out in the Contract. In case of litigation, the judge or the arbitrator is obliged to apply the law specified under the Contract.

Compliance with the applicable laws of the countries where the development Project is implemented: the Contractor must observe and comply with the rules of each of the countries where the work is being effectively performed, for instance in respect of safety or social laws. On the same development Project, you could have construction work performed in Abu-Dhabi, in Iran and in France. This principle is different from the choice of the law applicable under the Contract.

Arbitration: It is used when the Operator and the Contractor cannot settle amicably a dispute and one or both parties decide to go to arbitration. It is recommended that the selected arbitrators shall judge and settle the dispute according to the law applicable under the Contract and not in equity. Of course, we do recommend, as far as possible, to avoid arbitration, as it is generally a long and costly process (although it is generally simpler than the judicial procedure). It is sometimes possible to use the conciliation with an expert before going to arbitration.

The main types of arbitration clauses are ICC (International Chamber of Commerce) or UNCITRAL (United Nations).

As an example, we can give an arbitration case where the Oil Company lost in arbitration due to too many interferences in the day to day work of the Contractor; the arbitrators decided that the Contract between the Operator and the Contractor was no longer a Lump Sum Contract, but should be considered as a fully reimbursable Contract.

⌒ Independent Contractor

The independence of the Contractor in the performance of the work is essential. It could be expressed as follows: "the Contractor shall manage, control and direct the work as an independent contractor and perform all obligations and duties under the Contract at his own cost, risk and responsibility, in due compliance with the work time schedule and with the provisions of the Contract."

The Contractor has to be independent in order to be fully responsible for his obligations under the Contract. However, the Contractor's liability is not unlimited as it is usually admitted that the risks corresponding to consequential damages, such as loss of plant production or operating expenses, are borne by the Operator. A maximum limit of liability of the Contractor could also be fixed under the Contract: this value is equal to a percentage of the overall Contract price (up to a maximum of 100% of the Contract price).

The monitoring and control of the Operator must not disturb the "normal" performance of the work. It is essential that the Operator be not too directive towards the Contractor in the performance of the work.

The events originated by the Operator which affect the performance of the Work could be for example: an untimely delivery of equipment and materials supplied by the Operator to the Contractor or a large number of interferences in the day-to-day Contractor's work by the Operator.

⌒ *Work Time Schedule*

The Operator has to continuously monitor and control that the performance of the Contractor is made in accordance with the Contract work time schedule without incurring any "critical delay". A "critical delay" is a delay affecting the critical path of the work time schedule and, if not corrected, the Plant completion date.

Compliance with the contractual completion date is essential for the Operator as any delay in the completion of the work and in the transfer of the Plant by the Contractor to the Operator could affect the economics of the Project due to:

- Loss of Plant production.
- Extended financial charges.

The Contract documents usually include:

- The reference "work time schedule".
- A number of reference key dates taken from this schedule in relation with the performance of significant events in the performance of the work.

This time schedule and key dates are consistent with the mutually agreed work completion date and plant transfer to the Operator.

An appropriate monitoring of the progress of the work requires that the Contractor:

- Periodically reports the rate of achieved progress of the different parts of the work, to allow their checking against the corresponding reference "S curves" (curve used for the monitoring of the progress result).
- Immediately informs the Operator of any "critical delay".

A "critical delay" observed during the performance of the work may:
- Either only reflect a temporary difficulty which can be quickly corrected.
- Or lead to a limited delay in the completion of the work, slightly affecting the completion date.
- Or be the first indication of a loss of control on the work time schedule, which would significantly affect the completion date.

The risk of the third case requires the implementation of appropriate corrective measures:
- To oblige the Contractor to consider the work time schedule as an essential priority.
- To give the incentive and persuade the Contractor to take all necessary remedial measures to promptly correct any "critical delay".

These corrective measures may successively include:
- The obligation for the Contractor to quickly mobilise additional means.
- The take-over by the Operator of the management of the Contract.
- Ultimately, the termination of the Contract and the replacement of the existing Contractor, at his expense, by another contractor.

The Contract provisions must clearly refer to the "essential character" of time in the performance of the Contract.

To cover this situation, the English law has developed the concept of "time is of the essence". This concept takes its origin in the 19th century, when a tailor failed to timely deliver a wedding dress to a bride. The dress furnished after the marriage was therefore of no use.

The Court ruled that:
- The bride didn't have to pay the dress.
- In addition, the tailor had to pay significant damages to the bride as a compensation for the failure to deliver the dress on time.

The concept of "time is of the essence" means that any "critical delay", if not promptly corrected, might ultimately lead up to the termination of the Contract and the payment of damages by the failing party.

Liquidated damages for delay: In spite of all the corrective measures taken during the performance of the work to recover "critical delays", the Contractor may not complete the work by the due completion date. In this case, it is usually considered, as a compensation for the Operator for the corresponding delay, the application of "liquidated damages" to be paid by the Contractor. The amount of "liquidated damages" is defined by contract and is calculated:
- Proportionally to the actual duration of the delay.
- Up to a maximum amount (commonly taken at 10% of the overall Contract price).

The liquidated damages are applied after the completion of the work, to allow the assessment of the actual delay.

With the application of liquidated damages:
- The Operator needn't prove the damage due to the delay, but only refer to the actual completion date of the Plant by comparison with the contractual date;
- The application of liquidated damages generally avoids going to arbitration or to court where the Operator would have to establish the actual damage resulting from the delay and the associated costs.

The liquidated damages are the contractual tool used to compensate the Operator for the consequences of a limited delay in the work completion date.

They are not sufficient to deal with the occurrence of a significant delay in the performance of the Contractor.

In respect of the work time schedule, we do recommend that the Contract expressly incorporate the provisions related to correction of delay.

⌒ Contract Price

The choice of the type of the remuneration of the Contractor is an essential feature of the Contract: it strongly influences the Contract conditions. A fully reimbursable Contract will not need as constraining conditions as a lump sum Contract. There are several ways to compensate the Contractor for his work.

The remuneration of the Contractor for the performance of his work can be:
- Either on a reimbursable basis: open book basis, time rate basis with or without target, cost plus fee basis.
- Or unit price basis applied to bill of quantities.

- Or on fixed price basis (lump sum).
- Or a combination of the above.

The Operator can incorporate in the Contractor's remuneration positive or negative incentives to increase his motivation:

- Target man-hours: should the number of man-hours performed by the Contractor exceed the target, the man-hours in excess will be paid at a reduced rate.
- No-claim bonus/Surcharge: should the Contractor perform the work with a number of man-hours below the target, he will receive a bonus; on the contrary, he will be penalized with a surcharge should the number of man-hours exceed the target.
- Liquidated damages for delay, Plant or equipment performance: the Contractor will have to pay to the Operator a certain amount of money, as fixed under the Contract, in case of delayed performance or in case of insufficient performance of the Plant or of the equipment.

Invoicing and payment procedure: The timely payment of the invoices of the Contractor is an essential obligation of the Operator: failure of the Operator to perform such payment may entitle the Contractor to suspend the work and at the extreme to terminate the Contract.

The invoicing payment procedure in development contracts depends on the type of remuneration of the Contract.

The procedure can be:

- On a "progress basis", where the Contractor is usually paid every month on the basis of the progress of the work which has been assessed as achieved during the month.
- On a "milestone basis", consisting in the payment by the Operator of a number of successive instalments against the corresponding achievement by the Contractor of one or several significant events as defined in the Contract.
- On a "time basis" where the Contractor is paid on the basis of the man-hours performed by his personnel and equipment during the month applied to the respective contractual rates.
- On a "bill of quantities basis" where the Contractor is paid on the basis of the quantities of materials installed during the month applied to the relevant rates.

⌒ Variation Order

A Variation Order (sometimes called change order) is an amendment to the Contract mutually agreed by the Operator and the Contractor during the performance of the work.

The interest of such contractual provision is to set rules for the adjustment of the Contract price and time schedule in case of an event affecting the performance of the work.

The usual purpose of this provision is:

- To allow the Operator to request modifications to the Plant definition.
- To take into account the change in the conditions of the performance of the work, due to the Operator or to Force Majeure.

The issuance of a Variation Order accepted by the Operator and the Contractor usually implies the corresponding modification of:

- The Contract time schedule.
- The Contract price.

The Contractor may similarly request the issuance of a Variation Order in case he considers that modifications have occurred to the Contract scope or to the Contract conditions.

The Variation Order provision is not intended to allow the modification of the "general purpose" of the Contract. For instance, should the Operator wish to build a completely different Plant, this would justify another contract.

The evaluation of the price of the Variation Orders may be performed according to several methods:

- A lump sum proposed by the Contractor and accepted by the Operator.
- The application of the unit rates mentioned in the Contract to the additional equipment and materials quantities installed by the Contractor for the performance of the Variation Order.
- The application of the time rates mentioned in the Contract to the time spent by the Contractor's personnel for the realization of the Variation Order.
- On "cost plus fee basis", i.e. the reimbursement of the actual expenses incurred by the Contractor in the performance of the Variation Order increased by the application of the fee mentioned in the Contract to the amount of these expenses.

In case the Variation Order modifies the critical path, the Contractor is entitled to the corresponding extension of the Work Time Schedule. The Variation Order is usually formalized by a mutually signed document.

⌒ Force Majeure and Hardship

Force Majeure events include:
- Natural disasters, such as violent storms, cyclones, earthquakes, tidal waves and floods.
- Man-originated events, such as: wars declared or not, riots and revolutions, national strikes and lockouts.

Hardship events are generally events with economic or financial impact, which have not been anticipated at the conclusion of the Contract and could affect the viability of such Contract: substantial increase in oil prices or in wages, substantial compulsory legal reduction of working hours…

In the case of Force Majeure, the performance of one party becomes impossible due to the occurrence of the exceptional event. In respect of Hardship, the performance of one party has only become more onerous, but not impossible.

The criteria generally required for Force Majeure events are:
- Unforeseeable: events which can be foreseen at the time of conclusion of the Contract cannot excuse the non-performance of an obligation.
- Insurmountable: there may be lack of an irresistible or insurmountable obstacle if a party has itself, either wilfully or by negligence, caused the impossibility to occur.
- Out of control: there is no Force Majeure if the event triggering the impossibility is not external to the party seeking relief, but is related to its own sphere of power or influence.

It should be noted that under English law, it is not required that the three criteria mentioned above be met; the English law refers mainly to "out of control". From the Operator point of view, it is obviously better to refer to the three criteria to be met simultaneously.

We can give an example related to an Iranian Project: there were serious riots among the workforce (around 11 000 people) of the main Contractor between workers of different nationalities. This delayed the good performance of the work. This was not considered as a case of Force Majeure due to the fact that the Contractor should have had control upon his own workforce.

On the opposite, a rocket damaged an offshore platform located in Abu-Dhabi during the first war between Iran and Irak: this was considered as a Force Majeure case because the event was not really foreseeable, was out of control and was insurmountable.

The grounds of relief, in case of Force Majeure, are as follows:
- The failing party is relieved from contractual sanctions.
- The time for performance is postponed for the period of Force Majeure as may be reasonable, but each party bears its own additional expenses.
- Either party shall be entitled to terminate the Contract if the above period of Force Majeure exceeds duration of 3 or 4 months.

Hardship clause: Should the occurrence of events not contemplated by the parties fundamentally alter the economic or financial balance of the Contract, thereby placing an excessive burden on one of the parties in the performance of the Contract, that party may proceed as follows:
- The party shall make a request for revision within a reasonable time from the moment it becomes aware of the event and of its effects on the economy of the Contract and will indicate the grounds on which it is based.
- The parties shall then consult one another with a view to revising the Contract on an equitable basis, in order to ensure that neither party suffers from an excessive prejudice.

We do not specifically recommend, especially if you are on the Operator's side, to incorporate a "hardship clause" in a Contract: the substance of such a clause is highly subjective and its application could quickly lead to a potential conflict between the parties.

⌒ Warranty period

The acceptance of the Work and/or the Plant by the Operator is made through:
- The inspection and control of the conformity of the Work during its performance.
- The Provisional Acceptance of the Work and/or the Plant, issued when the Work and/or the Plant is completed.
- The warranty period after such Provisional Acceptance with the Plant under operating conditions (the Plant is operated by the Operator for production purpose).

- The Final Acceptance of the Work and/or the Plant at the end of the warranty period.

The warranty period is the period starting at the effective date of the provisional acceptance and during which the Contractor guarantees the good functioning and the performance of the Plant up to the Final Acceptance.

The Contractor commits himself to repair any defect that could appear during this period. This guarantee does not cover defects associated with the improper operating of the Operator or with normal wear and tear. The duration of the warranty period is commonly fixed from twelve to twenty-four months. At the end of the warranty period, the Operator issues a final acceptance certificate.

HSE objectives and human dimension

◎ What does HSE mean?

Health, Safety and Environment of course, as well as Hygiene and Security.

Some Companies separate "Environment business" from others. It is advisable due to the complexity of the problems. However at the Project level it is realistic to keep all HSE disciplines within a single point of responsibility.

How shall we differentiate Safety from Security? Security means all the actions able to keep properties information and data of the Company, or of the Project, safe from external predators or from lack of reliability of certain specific equipment. Safety is of course related to people (from the Project and from outside) and therefore to the material and equipment, including temporary facilities.

◎ The main objective of the Project

HSE is today the one main objective of the Project team. In order to achieve the expected results, the HSE policy is implemented from the very early stage of the Project. During the Pre-Project review the targets are defined and environmental assessment is evaluated. The Base line survey is performed, and will be used as reference.

We have already introduced some of the key words and actions; we can now enter into some more details:

- Hazid: Hazard Identification. One of the very first studies performed during the Feasibility Study period. A good knowledge of the Site is necessary.

During the last 30 years the list of potential hazards has been drastically increased. In addition to traditional earthquake matters or flooding risk by a river, it is necessary to consider many other causes: road impact, earth sliding (mud), terrorism, extremely bad weather, impact from other Plants, etc.

We can make reference to a disaster occurring in France some 50 years ago (1959). The Malpasset dam, built near the Frejus town, was suddenly destroyed, after heavy rains, by the rupture of the left connection between the concrete of the dam and the rocks.

Fig. 7.1a,b – Malpasset (Source : ecolo.org)

The town of Frejus was submerged within a few minutes and 423 persons were killed. In fact the water level of the dam could not be controlled due to the construction of a down stream bridge for the new motorway. Since this disaster, we haven't heard of any similar accident in France, but in other countries several dams were destroyed and the consequences were also terrible (India, Mekong River).

- Hazan: the Hazard Analysis purpose is to go over each of the identified hazards and to propose a review of the consequences and also the mitigation process. Such analysis is complex and it is often necessary to look at the statistic and at the deterministic approaches. Again this Hazan must be performed as soon as possible in the life of the Pre-Project. In some countries such as Europe it could lead to cancel the Project or at least to change its location.

Here is an example of a mitigation measure taken a few years ago to protect a billion dollar gas treatment Plant: heavy rain falling on the mountain nearby could cause sudden flooding. A concrete canal was therefore built to collect the water between the mountain and the Plant and to evacuate the water.

- Hazop: the exercise is very important and usually performed by very professional engineers from various disciplines. Very often the Hazop 1 is performed in parallel with the Project Technical Review N° 1. For a medium size Project the Hazop may need five to ten days. The "what if" approach will be used systematically. "What if the valve does not close? What if the valve does not open? What if the first firewater pump does not start...?" Some other Hazop will be performed later on.
- Safety Audits: during construction, regular Safety Audits will be performed. The procedure is essential as it is psychologically important to keep pressure on anyone working at Site. Tool Box should be checked at least on a weekly basis. Certificates for cranes, lifting devices should be controlled. Safety drills need to be organised, safety training as well.
- Safety reporting: every month, and often every week, the safety at Site is analyzed, the events are recorded and recommendations are made.

◎ Safety specific aspects

In many countries Safety is not part of culture. It is limited to boards such as "Safety First", but not to a specific training for all the personnel. Even for the Owner/Company the Safety behaviour has to be permanently updated.

Safety is not just the work of the Safety team. The Safety team is structured to help and inform, but the Safety responsibility relies on everyone.

Keeping in mind that very often the accident highest frequency is the transportation of personnel, improvement is possible.

But to fight at all levels the lack of safety concern is a difficult task: the individual behaviour is also one of the main causes of accidents. It is not just by wearing a helmet, gloves, safety shoes and glasses that Safety is successfully implemented. It is necessary for everyone to act safely by checking tools and materials, by defining adequate procedure, etc.

◎ One fundamental rule about Safety

- The ignition of a product can occur only if the 3 conditions are satisfied simultaneously (Fig. 7.2). If one of these 3 factors is cancelled, the ignition risk disappears. To stop a fire the same concept will apply, we can put out a fire with water or carbon dioxide ice: the fuel is separated from the oxygen (or air), and at the same time the Ignition Energy is taken out of the fuel.

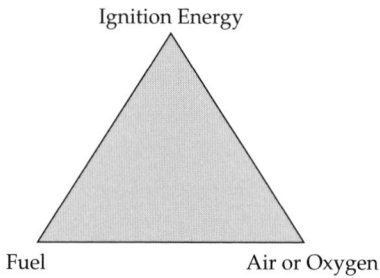

Fig. 7.2 –

- Some specific definitions apply to safety: LEL (Low Explosiveness Level), UEL (Upper Explosiveness Level).
- We must define at least one here. The Flash point: the minimum temperature at which a liquid releases enough vapour in order to obtain an ignitable atmosphere at its surface when using a flame.

A few other definitions:

- FMEA: Failure Method and Effects Analysis.
- LTIR – Lost Time Injury Rate.
- RCR – Recordable Cases.
- Flammable: a fluid capable to ignite easily, to burn intensively or to spread flame rapidly.
- Combustible liquid: any liquid with a flash point at or above 37.8°C.
  - Class II liquids – flash point at or above 37.8°C and below 60°C.
  - Class IIIA liquids – flash point at or above 60°C and below 93°C.
  - Class IIIB liquids – flash point at or above 93°C.
- Flammable liquid (Class I) flash point below 37.8°C.
- NFPA: National Fire Protection Association.

What about H2S? H2S is one of the worst hazards a project can face. The risk for humans as well as for the equipment and facilities is very high. Every year terrible accidents are faced here and there. When it seems that H2S could appear in the Plant either during early stage of production or later on, do not hesitate:

go and meet the safety engineer in charge, the corrosion specialist and any people who can quickly help you. At the level of 1 000 ppm, H2S will rapidly kill any people breathing such atmosphere.

◎ Security aspects

Security is also an important aspect. What do we mean? Securing information, securing equipment consists in applying all the rules and methods to avoid the loss of information or the destruction of equipment. Loss and destruction maybe caused by negligence, by misconduct, by loss of power, and also by terrorism. And sure we must not forget the traditional thief.

A few years ago, Security was limited to a fence, the control of the entries at Site and some routine inspections on the Site.

Today it can be much more complex: avoiding the spying of industrial data, protecting the Plant against modern terrorism may need dedicated personnel in the Company. The risk attached to the loss of data stored from computers is everyday more complex.

◎ Environmental aspects

Last but not least: the environment. Onshore and offshore, the environment is almost permanently at risk during construction. Discharging tubes or hydrocarbons, all the cutting of steel either from structural or piping, the dirty water, the surplus of concrete, welding rods, all the empty packing, storage, painting can, chemicals, and so on…

It is necessary to define specific rules and to oblige all parties to implement a very strict control of the rubbish and dirty water. Many companies today apply more severe environmental standards than the local law.

We remember visiting a Plant under construction in the South of Algeria: taking benefit of the transportation of many equipment and material by metallic containers, the main contractor started to align the dozens of containers to fence the site. Why not but what about the situation ten years later?

The demobilisation of the Plant is certainly one point to be addressed during the Pre-Project studies. How to turn back the Site into what it was before the Project?

Back to the middle of the 20th century no one could anticipate the problems and the cost of such site recovery. In the North Sea the dismantling of the platforms built for the Frigg Project required more than 2 years of pre-studies and the final cost is over 2 billions Euros. We can say that demobilisation is a real project by itself.

Fig. 7.3 – Nobody would like to be there!
(Source : *coordination-maree-noire.eu*)

Scheduling the Project – Preparing and follow-up

Scheduling is sometimes perceived as an easy task. It's not. Who has the time to analyze the overall tasks of the Project but a Planning engineer? Anyhow all the members of the Project team must have some very good knowledge of scheduling and of how to read a schedule. Anyone in the Project Management team should provide the planning group with all technical information. Sometimes a detail is perceived as having no value, in fact it maybe a very critical information for the schedule.

Do you want to build a Plant near the Red Sea? Look at the history of this geographical area: every beach, every border of the sea used to be full of personal mines. Can you guess the consequences for the project schedule?

Scheduling is one of the best ways to anticipate, and as we already mentioned: "anticipating" is really the basis of Project Management.

SOME KEY STEPS DURING THE PERFORMANCE OF THE PROJECT:

• **Site underground or underwater problem.** *Many difficulties and delays can happen due to the difficulties encountered in the subsoil.*

• **Onshore Rights of way.** *It is well known: no landowner wants a "dirty" Plant in his garden. Nothing but passing through a field, crossing roads or any properties means negotiation, compensation and… time.*

• **Basic Engineering.** *In order to establish clearly what the job is, it is necessary to study and prepare the key documents of the Projects. Reviewing all fundamental data, preparing the material balance and the utility balance, drafting a preliminary Plot Plan, setting the Process flow Diagram, etc. Many important tasks to define the Project clearly before doing any purchase or any construction.*

- **Selecting Long Lead Items (LLI), supplying them.** *For almost all projects, some equipment or material are more critical than others. For example: steel can be critical for a pipeline or for an offshore platform. A high voltage panel can also be critical. Therefore one way to limit the impact on such projects is to order those equipments with a long delivery time well in advance. But in order to do that safely it is important to complete a certain part of the Basic Engineering. By doing such an initial study, it is possible to assemble enough information to issue a request for proposals for such Long Lead Items. The work can be pursued up to the selection and eventually to the placing of the order.*

- **Spare Parts selection and supply.** *Spare parts are required at the end of the construction to replace some damaged material during transportation and/or site installation. Then spare parts are absolutely necessary as soon as the commissioning starts and during the start-up operation. It is therefore absolutely necessary to identify and purchase the spare parts well in advance to avoid interruption in the process.*

- **Financing.** *The financing of the Project must be available; it can be from an internal source or from an external one. If it is partly or totally from the banks, it will take time. Several months of negotiation and implementation will be needed. It will be faster if the funds are from internal sources, but some dossiers need to be prepared and transmitted in the Finance Division- a few weeks are necessary.*

- **Transportation/Transfer of large equipments.** *Today the purchase of material and equipment is organised worldwide. Consequently marine transportation and air transportation are one key task of the Project. We have difficulties to imagine the risk attached when the full Plant is transported over thousands of miles in just one go. For example the process Plant of the SnØvhit LNG Plant meant something like 33 000 Tons moving from Cadix (Spain) to the north of Norway.*

- **Offshore works if any.** *They are always critical due to the fact that the weather can be a severe constraint. The availability of marine equipment (pipe laying barge, derrick barge, tugs, transportation barge, logistic marine vessels) is also never granted. The cost of the offshore works: in average 3 times more compared to onshore works increases the critical aspects of all offshore works.*

But some onshore projects can also interface with the sea: each time a seawater pumping station is built. I remember a Project having to build such facilities and facing a lot of difficulties up to the time when two workers were killed during the work. The job was finally analyzed in details to avoid repeating such an accident. (MD)

- **Main on shore lifting.** *It is not just a coincidence: one of the very few French companies still involved in general industrial construction is a company with an incomparable expertise in lifting heavy equipment even inside congested areas.*
- **Commissioning and test.** *Starting at the end of the construction and when the Plant has achieved the mechanical completion. This phase main objective is to check the Plant and verify that every piece of equipment is ready for start-up.*
- **As Built drawings.......................and many others.**

The various methods

Bar charts and Critical Path

The critical path of a Project or a sub-Project is the sequence of tasks that do not have any time flexibility and whose overall duration cannot be reduced. Identifying the critical path is important but anticipating potential future critical path is fundamental. Scheduling helps defining the manpower necessary for the project performance.

The Bar charts is the usual method **to present** a schedule inside the Project team or to the Management or to the Client. The CPM or PERT is a logical approach where the fundamental rule is that tasks cannot just be done in parallel. There are predecessors and successors. I.e. you must pour the concrete foundation before setting equipment on top of it.

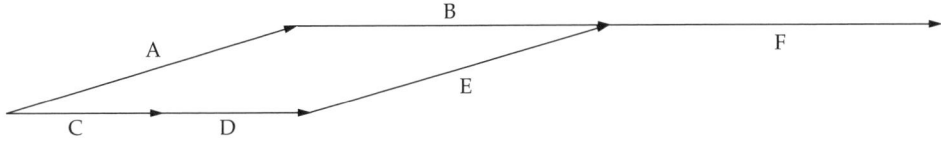

Fig. 8.1 –

From this very simple diagram, we can see that B is following A, E following D and D following C. In fact it means that B cannot start before the end of A and F cannot start before the completion of B and E.

The logic diagram must be established for all multi disciplines projects and as soon as the cost of the Project exceeds the 5 MUS$ budget. The main difficulty is that a real sequence as indicated here above is not easy to be identified for an industrial project: unless you go into a very detailed analysis many tasks can be performed in parallel at the macroscopic level.

If your project is a simple house, you can easily understand that you cannot start the electricity work before completing the walls and the roof. But what about the piping? Most architects will agree that it is better to install the piping before the electrical network.

◎ Collecting the necessary information

The scheduling cannot be done without a good knowledge of the Project. The various units, the main equipment list, the Plot Plan, the Process Flow diagrams, the main characteristics of all parts of the Plant, the Contractual Strategy… every project document is useful for the Planning Engineer. For the estimating engineer, it is also necessary to collect information about:

- The manpower productivity, necessary to evaluate the duration of the work at Site,
- The situation of the market, for example for the steel, the valves, the electrical cables… to evaluate the delivery time for equipment and material.
- The competition for all main works.

We can call such collecting phase, the "Analysis Phase". For an important project, a team must work together to analyze all data.

◎ Scheduling by Company and by contractors (Fig. 8.2)

The second phase: preparing the schedule is more like a synthesis phase. The Planning Engineer having all information in hand must place all the activities of the project. Very often a target date is defined at the top management level: production must start on or before a certain date.

The Company (we mean here the Client) is preparing the first planning for the future Project, in order to evaluate a realistic date for completion.

Scheduling is also performed by contractors and suppliers and reviewed by the Company Project team.

Scheduling the Project – Preparing and follow-up 79

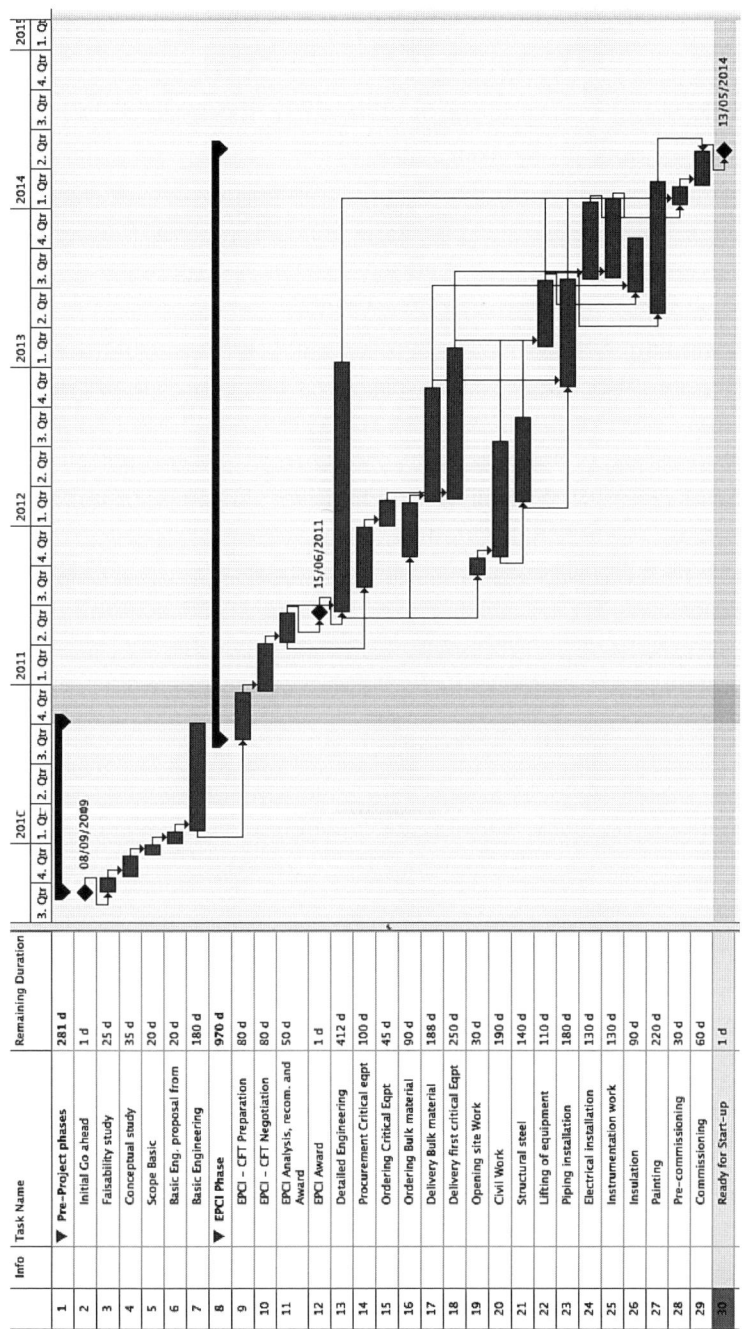

Fig. 8.2 – Typical schedule for an onshore project.

◎ Follow-up

The follow-up of the progress is permanently done with a special synthesis on a monthly basis. First it is crazy to try to calculate the progress of every thousand tasks part of the project one by one from 0% to 100%.

It is necessary for each discipline to identify the relevant physical quantity. But for engineering, procurement, management and others we cannot really identify physical quantity. That is why we will use conventional progress. One way to calculate the progress is explained by Fig. 8.3.

For physical quantities, we can use the progress by measuring the work done with physical quantities, but not money.

For example:
- M3 for storage tanks.
- M2 for exchangers, air coolers.
- Tons for boilers.

For structure:
- Tons or m3.
- Tons for fabrication and for construction.
- Onshore Pipelines: ml.
- Offshore pipelines: inch/ml – m3 concrete – m2 of coating.
- For Civil Work: cum.
- For Buildings: m2 floor.
- For Piping: ton/m2 for piping – weight split between valves, fittings and pipes.
- For Lifting: ton.
- For Transportation: % equipment.

Scheduling the Project – Preparing and follow-up

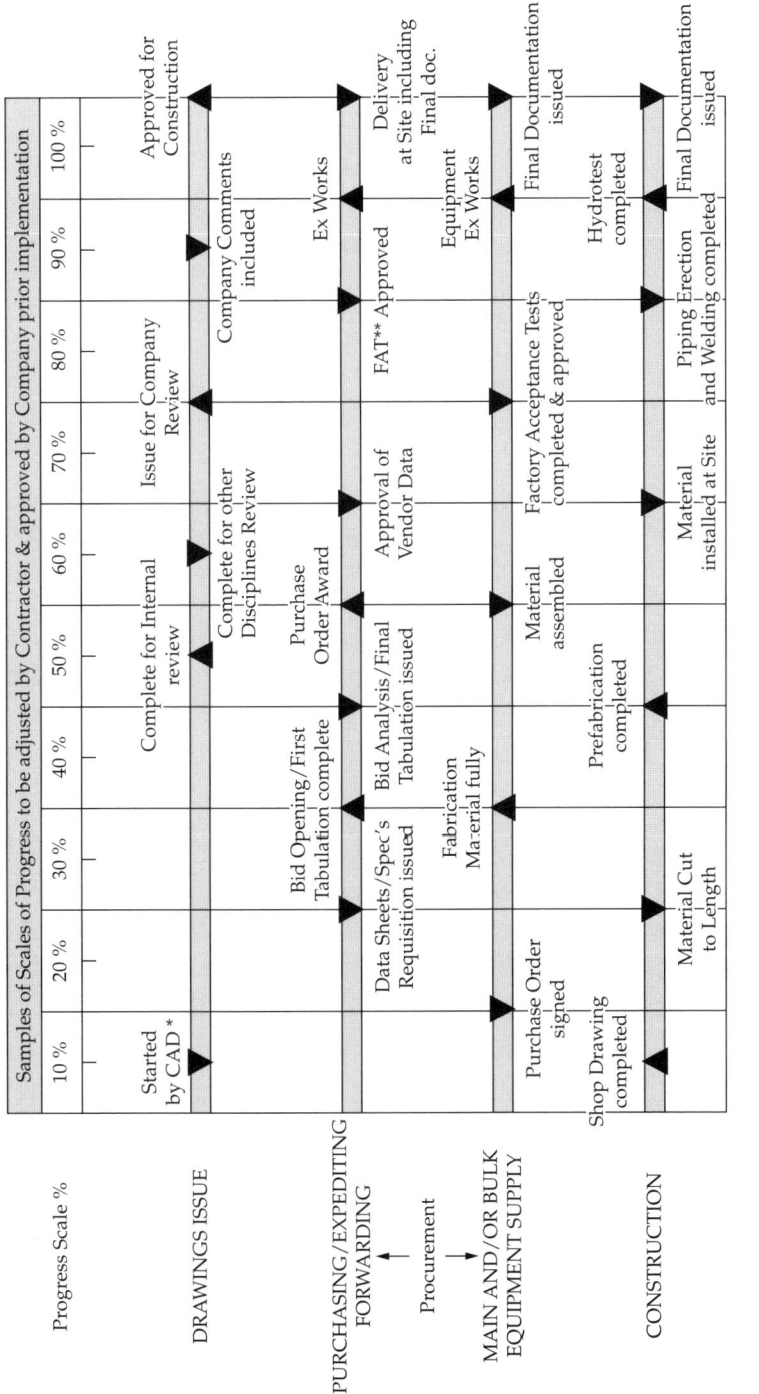

Fig. 8.3 – Planning and Progress control.

* CAD: Computer Assisted Drawing.
** FAT: Final Assistance Test.

Cost monitoring – the Work Breakdown Structure

◎ Cost and accounting

Cost monitoring has nothing to do with accounting. Starting from the cost estimate, the budget is approved. The cost estimate, the schedule and the cost reporting are based on the same Work Breakdown Structure (sample of WBS, Fig. 9.1).

◎ Cost monitoring: why?

- To coordinate the efforts of the Project team to keep the expenses and the commitments within the envelope agreed by the Management and the Partners.
- To ensure that the expected economic criteria of the Project will be met.
- To propose wording and clauses to be incorporated in the contract wording.
- To keep the same methodology across the Project and sub-Projects and to verify the good circulation of the information.
- To prepare the cash call with the Finance Division to ensure that the Project will have the funds to pay the contractors and suppliers.
- Project duration may increase the difficulty of the cost management (currency fluctuation-inflation…).
- Investments face more and more complexities and amounts often move towards very high figures.
- Several parties are involved (contractors, suppliers but also partners…).

Cost monitoring – the Work Breakdown Structure

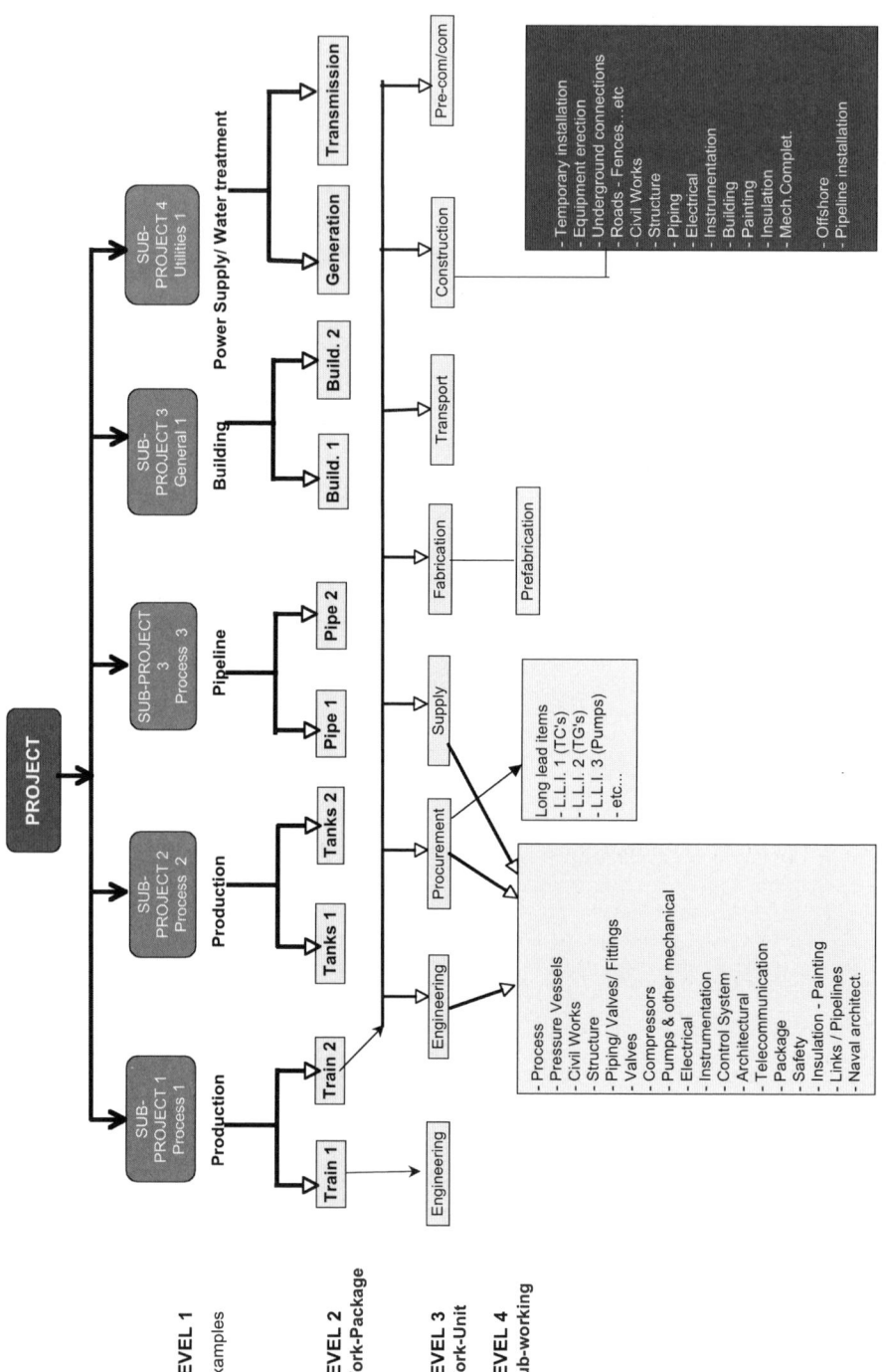

Fig. 9.1 – Work Breakdown Structure.

◎ Cost monitoring fundamentals

Cost monitoring means evaluating all potential deviation and keeping control of all remaining commitments to be placed. The Project will always start from a Budget.

- Once the Cost Estimate is ready and the Project agreed by the Operator and the Partners, the **Initial** Budget is finalised. Built in accordance with the WBS structure and the Project Schedule, this Budget will also be a Life Of Project (LOP) Budget.
- The accounting of a Company is normally based on a yearly basis, while a development Project must be followed along all its "Life". Project Commitments (see below) have to be followed up on a long-term basis.
- Initial Budget may require some adjustment during the Project Life: either for the transfer of a provision or a budget line from one Package to another, or the deletion/addition of one line. The Budget amended will become the "**Revised** Budget".
- Thus at the end of the Project, the final cost analysis will create the "Final Budget".

◎ Cost Forecast

- Cost Forecast is updated on a regular basis (monthly).
- Cost Forecast may anticipate on some not finalized commitment but Confidentiality must be addressed very cautiously.
- Cost Forecast may require a good knowledge of the technical matters.
- Cost Forecast may introduce some budget transfer (see Fig. 9.3).
- Provisions can only be used with the agreement of the Management and in line with the Management Agreement and the Financial Agreements.
- It is advisable not to use the Provisions during the early stage of the Project.

Budget evolution: we can see some ups and downs for the budget. The commitment curve jumps quickly due to the signing of main supply and main contracts. The expenses curve represents the value of the work done. We can call that curve the "expenditure incurred" (see progress measurement). The payments curve (we can also draw the curve for invoices) follows the shape of the expenses curve but 3 months later. This curve is very useful for the cash call of the Project.

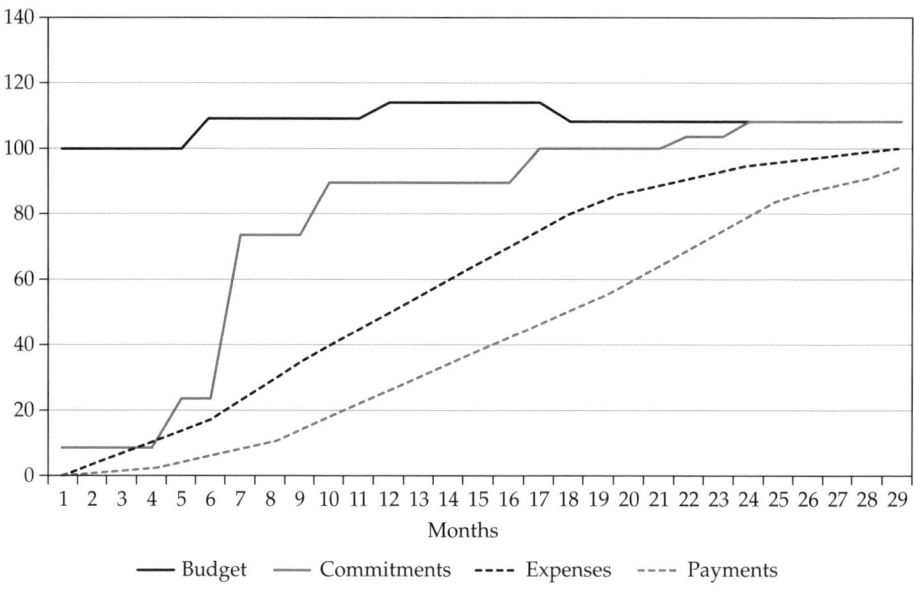

Fig. 9.2 – Cost evolution.

◎ Payment of contractors and suppliers

Payment of contractors and suppliers has to be organised fairly. It means a technical approval of the work done.

A contractor can be paid on the progress of the work done, or on specific milestones as defined in the contract.

There are pros and cons for both methods: A milestone is an event but the event must be precisely defined and must not be disputable when it happens. It is not always easy. For example if we select the "delivery of a main equipment to the Site" as a milestone, it looks fine at first glance, it is factual. But what about the documentation attached to this equipment? If the documentation is not delivered with the equipment, can we say that the delivery is fully performed?

Progress is significantly more difficult to implement but much less controversial. We have to take some examples, but in any case:

- Any Commitment must give reference to an invoicing procedure.
- Invoices must be prepared in accordance with the Commitment and submitted along the invoicing schedule as described in the Contract Agreement/Exhibits.

- Advance or down payment maybe considered.
- Invoices submitted in advance are not acceptable.
- Invoices disputed may or may not be partially processed.
- Any invoice movement is recorded.

◎ Reporting from the Project team

- Legal reporting (financial periodic reporting) is today mandatory.
- Fiscal reporting (depreciation…) is necessary for the finance division.
- Operational reporting specially for cash call with Company and Partners.
- Management reporting.
- Reporting to Partners.

◎ Cost monitoring day to day action

- Reviewing all progress of the work and the coherence with the contract terms.
- Addressing to the management any order/contract not budgeted.
- Reviewing all invoices received and the coherence with the contract terms.
- Reviewing on a permanent basis (at least every month) the estimate to complete and the final forecast.
- Managing the various provisions.
- Calculating the exchange rate impact, if any.
- Calculating the escalation formulae and price revision, if any.

◎ Change Orders

"Change Orders" are not an anomaly:
- The Project Cost Management team will monitor Change Orders as soon as they are identified.

- Change Orders being evaluated by the Contractor, such estimation will be checked with the assistance of the Cost Management team.
- Change Orders before being formalized and agreed by all Parties will only be introduced in the "Estimate to Complete" (Fig. 9.3, Column 6). Therefore a potential Change Order is just similar and part of the initial provision.
- Once the commitment is formally agreed, the corresponding amount is transferred to Column 2.

We cannot explain everything about negotiating "Change Orders" here (or "Variation Orders"). It may be fair to pay something more to a contractor or a supplier, it maybe crazy too. **Technically it has to be justified, contractually it must make sense.** But very often facts are driven by economics: on one side a contractor is trying to get as much profit as he can from the contract, on the other side the Owner/Company is committed to start production.

Unfortunately in many countries we face many difficulties to follow a change order procedure for administrative reasons and lack of delegation. Such policy does not protect the Project against some price increase. In fact it is the reverse. Any refusal by the Client to compensate for any additional work may push the party suffering from such situation to the extreme solution including court settlement, pushing back the completion date, reducing quality, etc.

Quarter Platform — TOTAL E&P XXXX — DATE: OCT 2004 — ALL IN k US$

CODE	DESCRIPTION	(1) APPROVED BUDGET	(2) SCOPE CHANGE	(3) BUDGET TRANSFER	(4) CURRENT BUDGET (1+2+3)	(5) COMMITMENT	(6) ESTIMATE to COMPLETE	(7) FORECAST (5+6)	(8) Variance (+/-) (7-4)	(9) Expenditure Incurred	(10) Approved Invoices
310	Long lead Items	2560	0	0	2560	2560	0	2560	0	2560	1250
320	Gen. Mngt & Serv	1200	0	0	1200	1200	0	1200	0	240	240
330	Engineering	2150	0	0	2150	2050	100	2150	0	410	410
340	Mal & Eqpt supply	12720	0	-250	12470	11890	580	12470	0	2600	2600
350	Onshore fab	6140	0	0	6140	5725	415	6140	0	573	573
360	Transportation	2200	0	0	2200	2125	75	2200	0	0	0
370	Offshore Works	5400	0	0	5400	5100	300	5400	0	0	0
380	Hook-up Wks	3600	0	-250	3350	2950	400	3350	0	0	0
380	Commissioning sup	800	0	0	800	600	200	800	0	0	0
390	Spare parts	300	0	0	300	300	0	300	0	0	0
3***	TOTAL	37070	0	-500	36570	34500	2070	36570	0	6383	5073

Fig. 9.3 –

Quality assurance and Quality control – During the Project and up to commissioning

Quality is not only governed by international standards. It is specific to the Project objectives: the reliability of a Plant is often a fundamental criterion for the successful economic conclusion of a Project. If you ask a computer chips manufacturer about the supply of power, he will tell you that 99,99% reliability is a must.

The Owner of the Plant, or the Operator acting on behalf of the Owner, is not able to check alone the quality for all the steps of the Project. The Operator will work out a *Quality insurance program* to manage all Quality procedures prepared by the contractors and the suppliers of the Project.

◎ Quality at the design level

The Company works with a referential: it means a set of documents including internal rules and general specifications applicable to all projects and all disciplines.

The project is then further defined by particular specifications for all equipment and the main elements of the Project. The project team will check the quality system proposed by contractors and suppliers, and will ask for regular reporting.

Quality for the design has been historically one of the most difficult tasks to achieve. The main reason is due to the necessary dread of all disciplines being part of the project: i.e. process piping layout must not interfere with structural steel or electrical cables layout. Who is going to check such interface and who is going to propose a solution? Today the computer (CAD 3D system) may help identifying the potential clashes, but the solution has to come from a very qualified technical coordinator. The technical coordinator is someone with at least 15 years experience who is able to evaluate quickly the relative importance of all disciplines.

Today due to the drastic price increases all around the world, any contractor or supplier tries as far as he is concerned to reduce its own upstream cost. Obviously this can have an impact on the quality. Therefore for a Buyer or a Client, if particular strict criteria for raw material and fabrication are not specified, the risk of obtaining poor quality is eventually important.

I was recently involved with an engineering company in a middle-east country. I cannot say that the quality of the work was poor, but I can say that supervising and controlling were absolutely necessary. The personnel involved were not under qualified but the turnover of personnel was certainly a main cause for creating quality problems at the engineering level.

◎ Quality and inspection during fabrication

Fabrication here means the work done at any workshop and/or fabrication site to perform a part of the Work.

An inspection program is prepared and the Company may decide to participate to some of the inspections, while suppliers and contractors will organize their own quality inspection and reporting.

For piping or pressure vessels the quality is checked by NDT (non destructive test) and often pressure tests, plus by all visual tests and, of course, by setting together the collection of material certificates.

Each Company has its own rules to supervise specific fabrication. At least for most of the fabrication, there is common concern to obtain good quality, and consequently to implement the relevant actions, for all rotating equipment, most valves – especially for special services valves – and for all lifting equipment.

Today and more specific to the Oil & Gas production, the deep sub sea equipment and material are technically critical.

But for many energy projects there are always common concern for pipelines and offshore structure. The explanation is almost obvious: if the selected steel is not of the expected quality and as specified by the technical specialists, the Project maybe jeopardized.

A very well-known example is the following: for the transportation of gas, it is fundamental to specify the material in accordance with the quality of the gas:
- If the gas is wet and contains carbon dioxide (CO_2), corrosion will occur and it is fundamental to deal with the point during the design phase: at least the "corrosion allowance" will be taking into account the corrosion risk.
- If the gas contains some H2S, the steel is not standard – NACE rule will certainly apply and an expert who will review all data and parameter including H2S partial pressure will determine the steel quality.

> *Example 5 – Test of gas compressors in a factory.*
>
> Some fifteen years ago, I was working for a gas project and the question was whether a full testing of the gas compressors at the factory plant was necessary or not. Today we could face the same problem. The accurate testing in a factory is not obvious: a large quantity of gas is required and all the utilities have to be there.
>
> Is it worth doing such a test? Well it depends: if the final destination of the compressor is offshore or maybe in a very remote area, what is the consequence of a non-identified problem occurring at the start-up? The repair may be just impossible or does require the mobilisation of a large maintenance team. The worst scenario will be to have to return the compressor to the factory and to lose production for weeks or months. Then the cost of the test at the factory may look very reasonable.

◎ Inspection during construction

The inspection of the Plant is performed almost everyday. Within the Project team, several supervisors will continuously check the work done and the conformity with the drawings.

Let us list some defaults which could lead to a very difficult situation later on and which can hardly be discovered when the work is just completed:
- Insufficient soil substitution, or done in an improper way.
- Bad drainage.

- Underground piping or underground cables laid down without good positioning.
- Concrete using salty water, rebars not properly covered by cement-concrete drying too quickly.
- Wrong selection of anchor bolts.
- Inside of drums, columns, reactors not properly checked, bad cleaning.
- Use of steel or pipe from a different grade or characteristics than the one specified, or lack of material certificate...

Consequently the supervision team has to work intensively from the AFC (Approved For Construction) drawings and has to visit the Site on a daily basis.

Formal meetings will be organised with the main contractor but also with sub-contractors, as necessary, on a weekly basis.

◎ Final checking: pre-commissioning and commissioning

Pre-commissioning and Commissioning allow the full checking of the Plant and its conformity with the design before start-up.

Such tasks can be started before the final site delivery of equipment and material. It means that at any Worksite, it could be worth starting pre-commissioning activities there. Qualified people are there, as well as testing equipment. It is also very often possible to start commissioning activities. This approach is very beneficial when a significant work package is performed.

The main differences between Pre-commissioning and Commissioning are the following:

- Pre-commissioning means static test and checking material and equipment against the drawings and the Vendors documentation.
- Commissioning means starting energizing the Plant, by connecting for example to the electrical network. The commissioning is usually performed by system. It means the Plant is divided into systems and sub-systems. A system is an independent part of the Plant and may include various disciplines such as piping, mechanical, instrumentation and electricity.

Once all and every sub-systems are commissioned, a certificate must be signed by all parties involved and of course by the Client and/or Owner of the Plant.

The construction phase – Relationships with suppliers and contractors

◎ Organizing the construction

The construction phase is organised during the engineering and the contracting phases. The Owner/Company has to be "present" on all fabrication and construction sites. "Present" means that at least one permanent representative is at any site where a part of the work is performed.

◎ Site selection

The final construction site will often require specific logistic facilities: road, harbour, accommodation, canteen, temporary offices and storage areas.

It may take two years to organize the final Site before starting the construction of the Plant.

- A local survey is first performed, then as before mentioned, an Environmental Base Study will be required if the site is virgin. The key parameters are identified: soil characteristics, weather forecast, altitude, existing construction, actual residents, rural activity…
- The temporary facilities are listed and quantified: offices, habitation, canteen, school, clinic, security, safety (fire fighting), telecommunication, power, general water and drinking water, sewage system, material and equipment storage areas. A plan has to be prepared and proposed.

- The entrances to the Site are analyzed in detail and the new routes, harbour, airport are evaluated. The construction of these facilities, if required, will proceed as soon as possible.
- The temporary facilities, listed above, are built.

Example 6 – When I started my career in the Oil & Gas industry, I had the chance to visit an island in a Middle East country where a huge Plant had been built and was under extension. Thousands of workers were there. Most of them were coming from Asian countries. I was amazed by the organisation to house all these people: rooms, offices, clinic, airport, harbour, police, custom, etc… The most surprising for me at that time was the clubhouse for the staff: a real British club. Since then many camps have been built and the rules to be followed are getting more stringent year after year. For example schools for children are part of the Project. In many countries it is moreover necessary to provide minimum wages for all workers working at the site, and at least, for the foreign manpower, to provide them once a year with a return ticket to their home country.

◎ Administrative matters

Many administrative matters are faced. Construction may be delayed or even cancelled by local constraints and it is important to anticipate and to evaluate the risk. Informing the local authority is part of the Owner's role.

Even if the construction is driven by a major contract, the Owner remains legally responsible and has to keep the control of the overall site organisation.

◎ Fabrication and prefabrication

We differentiate between "Fabrication" and "Construction".

"Construction" deals with the performance of all and any work at the final Site.

"Fabrication" deals with the performance of all works realized in a workshop or sometimes a yard to fabricate or prefabricate a part of the future Plant. For example very often we skid some equipment to ease the transportation and the installation at Site. Moreover the piping can be fabricated by making spools. An isometric can represent two or three spools. A Contractor within a dedicated area nearby the place where the equipment is fabricated does this.

Fabrication is often an advantage. The productivity in a workshop can be very good, and always better than at Site. Quality control is made easier. Interfaces with other contractors are nil. The Site cleaning is also simplified: less rubbish.

We do recommend fabricating or prefabricating as much as possible.

◎ Offshore specific scenario

Offshore, the situation is specific and an adequate safety procedure is put in place. Dredging, heavy transportation, lifting, scaffolding, are watched and specifically analyzed. The logistic is one essential aspect of the offshore Project.

First we know that for every productive employee working offshore, we need at least one more person to take care of the catering, security, medical, communication, transportation, etc. But other aspects of the logistic can sometimes be complex: how to deliver missing spare parts during the commissioning? You may need a chopper to be at Site within 24 hours. Then the cost of transportation can be equal or above the price of the spare parts.

◎ Performance of the construction – the discipline sequence

The normal way to follow the construction of the Project is to assess the progress by disciplines[1], we mean here:
- Site preparation.
- Civil Work.
- Underground network.
- Buildings.
- Installing equipment.
- Structural steel.
- Piping installation and connection to equipment.
- Electrical work and earthing.

---
1. See Fig. 8.3.

- Instrumentation work.
- Insulation.
- Painting.

There may be some other disciplines from time to time.

Fig. 11.1 – Civil work is often perceived as an easy discipline. It's not.

◎ At the end of the construction

The Commissioning Phase, final checking of the Project, is often organised starting from the engineering phase. We may differentiate between Pre-commissioning and Commissioning. The first one deals with static checking and conformity with drawings.

The second one is more dynamic and utilities (such as air, electricity...) will be used, but the raw product is not yet introduced.

The start-up will include the introduction of the product to be treated, and, for a refinery, a chemical Plant, a gas Plant or an oil/gas production Plant, it means oil and/or gas. From the start-up on, the Project is below the responsibility of the Operator. The risk of the Plant is transferred out of the Project team.

◎ Relationships with suppliers and contractors

Relationships with suppliers and contractors are not easy. Conflicts may happen, the sooner a solution is agreed, the better it is. In this Chapter we have just seen that many disciplines and many different skills are needed. This can only be achieved by getting expertise from everywhere. But various entities may have different objectives or at least may consider the same problem in a different way. So discussions are very useful, not to say absolutely necessary.

Relationships with suppliers and contractors are today governed by mutual respect. It is out of the question to squeeze suppliers and contractors. On the long term, as a client, we expect to keep competition between contractors. Forcing contractors and suppliers to reduce their prices, to squeeze delay or to get the payment of their invoices late, may have very dramatic consequences… even for the Operator.

Naturally, as we have seen in various chapters, it is part of the Project Management to check quality, schedule and safety. Checking and reporting are important aspects but they are the basis for the next step: correction. Correction is achieved by underlining the wrong attitude and manners and most of the time it means corrective human actions: safety training and drill, additional personnel, changing organisation changing people, etc.

A few years ago and even today, in several structures and countries, the cultural management has been driven by conflicts and imposed decisions (we are not here speaking about disputes). A legal conflict however is rarely beneficial to the Project development. Building a legal/contractual dossier will require a full team and hundreds of thousands dollars. As a consequence, a legal dispute will last for years, the cost will be huge and all Parties will be losing energy and time.

Can we really say and believe that Project Management can be based on conflicts?

The rule is then to negotiate discrepancies between owner and contractors/suppliers on the measurement of the performance (quality, quantity and schedule) as soon as possible. We are not saying that the Project Manager has to be weak or to always agree with Suppliers and Contractors. He must be fair and sanction errors and mismanagement, and be flexible enough when the wrongs are divided. Definitely and as mentioned here above, if the necessary corrections are not implemented by the Contractor/Supplier, the Client has to implement sanctions himself.

> *Example 7 – Errors in Client's information.*
>
> We can take an example with drawings and/or specifications supplied by the Client. The contract provides for the Contractor/Supplier to check such information and to inform the Client for errors and discrepancies. But very often such information is really from the Client, it could even be confidential data. Can we blame the Contractor for an error from the Client? In accordance with the contract, the answer is yes… but in fact, it is a case-by-case scenario and a very tough attitude from the Client could be perceived as the starting point for a conflict.

◎ Relationships with local authorities

Relationships with local authorities require a lot of knowledge and diplomacy.

The Project Manager's qualification is necessary, his human behaviour is essential.

Moreover local authorities are not often easy to manage. First the regulation has to be analyzed and understood, then the impact on the Project Development evaluated. But regulation is one aspect; the way to implement is another. It means human interferences.

> *Example 8 – Approval of a Project.*
>
> *The Project was in Asia and at that time the process to approve an Energy Project had to go through the specific department of Mineral resources (moving from one Ministry to the other). As we will see, AFE (Authorisation for Expenditure) was the process (as it is still today for many projects). So I prepared the relevant documents. The critical point was: I had to present only a part of the planned expenses, qualified as "investment"; the other part was qualified as maintenance (not subject to approval). Very quickly we understood that sending paper was not enough. I had to go and visit the department. I spent half a day introducing myself to the various services and departments and then it was time to explain my proposal in a meeting. The local staff of the Company of course supported me. I soon understood that the only problem was neither the amount of the investment, nor the reason to invest. In fact "I was just the problem: it was the very first time that I was going through the Ministry. I was a new guy. "Can we trust this new guy? Certainly not if all he can do is write papers. But maybe we can, since he comes and spends time with us, since he answers questions. Yes indeed, we can trust him now!" (MD)*

Training for project and operation – When to start?

◎ Training for the project team itself

Is it reasonable to start working in a country without any understanding of the local language or any basic knowledge of the specificity of the country? It is certainly too complex, and it will take too long, to train all the project team to be fluent with the local language. But having some basic training in order to be perceived as a polite person is fundamental. The "dos and don'ts" can be summarized in a small book. One week training language could be of a great help for anyone arriving for the first time in the country. Better if it is more.

> *Example 9 – Trans-cultural weekend.*
>
> *When I was called to work for the second time in Thailand, all the Company staff was invited to attend a specific weekend in a very nice place outside Bangkok. The spouses (or husbands, or boy/girl friends) were also invited. A specific Thai teacher having good knowledge of western behaviour organised the course. It was very useful and interesting as well for European as for Thai staff. We learnt about our differences and also our prejudices. Just for example: in Thailand you don't invite your colleagues at home. Should your home be more luxurious that the one of your colleague, you will offend him/her. If it is less luxurious, you will be the offended person. (MD)*

◎ Training for future operating personnel and maintenance team

Each Project is specific, but the training of future operators will start well "in advance", depending on the complexity of the Plant and the level of candidates.

Modern Plants are often equipped with sophisticated control systems and complex software.

The Operating Philosophy prepared during the FEED/Basic engineering defines the number of future operators. Today the trend is to try to reduce the number of operators and maintenance people but this is only possible when the local industry is really organised.

The training period could be six months, could be 2 years. Recruitment and pre-selection phase may already take several months. The Project language often affects training.

Interfaces between commissioning and future operators are thoroughly analyzed.

◎ Organising a selection of operating personnel and building a training school

We could make a list of countries where a training school had to be organised in a very systematic way by professionals.

An experienced operator may be available on the market but it is not always the case. If you work in a quite underdeveloped country, it may be difficult to find qualified production staff. If you work in a reasonably developed country, you may have to fight the competition.

For the Energy sector, the manpower demand remains very high worldwide. I met some western people able to double their monthly salary by accepting to work in a foreign country.

It is therefore necessary to organize a recruitment campaign. If you are part of a large international Company, it may be relatively easy. Announcement in some well-known local newspapers may be very successful and hundreds of résumés and letters of interest will come up.

The recruitment may be more difficult for a small company. The fact that the company is not supported by a local image yet is certainly a serious handicap. Recruitment will have to be supported by a special agency.

Then setting the training school:

- Finding an adequate building.
- Mobilizing teachers.
- Deciding in detail on the training program.
- Deciding on the intermediate and final examination.
- Helping the students to come and stay at the school, etc.

It is easy to understand that such task cannot be achieved but in at least two years.

Many important Projects are rising today all over the world and often in countries under development. Therefore there is a lack of manpower. Training is definitively one of the keys of success.

But even in developed countries, the problem may come up: recently a well-known international steel fabricator has been trying to resume one of its main production unit stopped a few months before, due to the international economic crisis in 2008. But where could one find new operators? Many of them are under a retirement program.

The baby boom is over!

Summing-up and conclusion

A Project is a human task. Without human participation nothing will happen. The Project should never be built or organised against the people or population living nearby. However some projects are controversial on that matter: a chemical Plant, a nuclear Plant, even a hydraulic dam can be considered as dangerous for neighbouring people. That is definitely why the initial phase of the Project deals with Impact assessment and risks evaluation.

One of the main difficulties today is to obtain all approvals for a new industrial project to be set in the countryside or nearby an urbanised area.

In the countryside the main concern is the impact on the environment, while in an urbanised area the concern is with the safety of the people.

◎ A Project is always unique

There are never two identical projects. You can learn from a Project but the next one, even for the same purpose, in the same country, will be different. The soil may be different, the schedule may be different, and the technology and the equipment will be different.

The Project team is the centre of all decisions (important ones but also the many day-to-day necessary decisions) and all information should go through it.

The Project Manager organizes the Project as per the Company rules but he will be the captain of a team and will distribute the work. He has to remain open to all comments.

As we have previously seen, his overall role and behaviour consist in:
- Leadership over the team members.
- Intellectual rigor. He never compromises with the truth.
- Ability to make the right decision at the right time.
- Having a good sense of anticipation.
- Being technically competent due to a worldwide experience.
- Good common sense.
- Ability to communicate.
- Strong personality.

Finally we can sum up:

◎ What is a "good" successful Project?

- A Project in line with the economic Company's criteria: Capex, Opex, schedule (Plant start-up date), production (build-up schedule), and consequently delivering to the Owners the expected revenues along the Project Life. We have underlined the basic rule: anticipating and forecasting.
- A Project achieving high standards of safety (very low accident record) during the development and preparing the Plant and the operators along the same strategy for the production phase. Safety has to be a permanent concern at all levels and for all disciplines.
- A Project complying with environmental impact assessment, but also organised specially during the construction in such a way that environment should not be at risk and,
- A Project achieving easily maintainable units at a reasonable cost. This aspect is only achieved when the project team and the production people have accumulated a large experience and are closely working together.

A Project is unique, but the result of all projects can be measured against similar criteria, and a good project certainly matches the specific goal of the Company. A successful Project is for sure a project where the participants can feel proud and satisfied as the achievement is in line with everyone's expectation.

We can terminate by a nice story to be remembered! *Charles Peguy's story*: On a very hot summer day in the Middle Ages, one traveller met a first guy carving stones for the construction of a cathedral. He asked him how he felt. He answered: "I am forced to work as a slave under this heat doing something I don't like. I hate it". The traveller went further and met a second guy. This one at the same question answered: "This job is difficult in this hot day, but this allows me to feed my family and I have to accept it". The traveller continued his walk and met a third guy who looked enthusiastic, contrarily to the first two. He repeated his question. That one said nothing but: "I am building a cathedral."

Building a Plant either onshore or offshore is a fascinating and exciting work... which could affect you for the rest of your life! The stakes and the risks are enormous, not only for the Project team, but for all the persons and entities involved in such a project.

A development project could involve thousands of workers; the resulting production may also contribute to the improvement of the lives of many more people for several decades! The Project team is not doing such a task just to satisfy themselves or to satisfy the "boss". It means work for thousands people, and most of the time additional production will be delivered by the new project for 15 to 20 years or more.

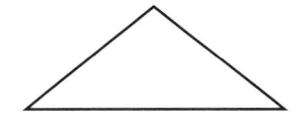

Key definitions

So it is time to review some of the key definitions used for international projects.

AFE: Authorisation for expenditure – *a formal document, often requested by the local authorities to approve proceeding with the investment.*

Approval: *for a budget, a partnership, a contract, a document or any commitment, approval cannot be given but by the right level of the management.* So we can say that "Approval" requires the right level of authority.

Basic Engineering: phase of the project where the main key documents are prepared in order to have them ready for the EPSC contract or any construction contract.

Budget: initial, revised, final – it has to be approved by the Company Management and the Partners – *the budget is prepared from the cost estimate. It may be revised from time to time for example once the main commitments are well-known and evaluated, but also when a change of scope is decided and agreed.*

CFT/RFQ: Call for Tender – Request For Quotation. The wording ITB (Invitation To Bid) is also used.

COD: Commercial Operation Date – Also to be considered: Mechanical Completion, Reception…

Commissioning: checking the Plant with fluids and energy. Performing all tests with adequate piping and equipment when and as necessary. *See also below Pre-commissioning, an activity performed just before the Commissioning.*

Commitment: any order, contract, change order, instruction, either formal or informal, creating relationships between two or several parties.

Company: the company in charge of organising and setting the development of the Project. *The Company can be also named "Operator" when representing a joint venture of companies. "Operator" can also be one of the "Owners" of the Plant.*

Contractor: a party in charge of performing part of the work, and often in charge (partly or fully) of the supply of equipment and material.

CPM: critical path method, also called PERT (Project Evaluation and Review Technique), *the method to evaluate and draw the schedule of the Project by following a logic analysis.*

Detailed Engineering: in general performed by the EPSC contractor, or the Engineer and finishing with the as-built drawings. *To differentiate from the Basic Engineering, the detailed engineering means all the study and drawings necessary for the construction of the Plant.*

Engineer: either an independent party or an entity part of the EPSCC's contractor group.

EPSC or EPSCC Contract: Engineering, Procurement and Supply, Construction and (Commissioning) contract. *This type of contract is of course used today mainly for large projects. As an order of magnitude projects over 50/70 millions $. We can also have EPCI: Engineering, Procurement, Construction and Installation, which may be different.*

Guarantee: used in particular for performance of the Plant. The Owner may ask a Contractor to guarantee the performances attached to a part of the Plant.

Hand Over: procedure taking place at the end of the commissioning.

Liquidated Damages: sum due by any supplier/contractor when the performance cannot be met or a non-excused delay is faced.

LLI: Long Lead Items and critical on the Work Time Schedule, sometimes separate orders (turbines, compressors…) can be placed by the Owner or the Manager before EPC award.

Lump Sum Contract: a contract based on an agreed price representing the overall value of the Project.

Manager: maybe the representative of the Owner. *In many countries the Owner has to be a private or public entity from the country, therefore the foreign entity may act as a Manager.*

Milestone: an event in the life of the Project, very clearly identified and having some importance for the overall progress of the Project. A date of completion or delivery is often attached to the milestone.

Operator: the party in charge of operating the Plant after the construction *(maybe different from the initial leader of the development)*.

Owner: can be a sole Company or a Joint Venture… the Party in charge of funding the Project and legally owning the Project Property. *It can be slightly different when the local entity is just a subsidiary but the funding company is in fact the mother Company.*

Plant: the objective of the Project; it can be geographically disseminated. It will be what we can see at the end of the Project.

Prefabrication or Fabrication is the part of the Work performed outside the SITE in a specific place: workshop of one of the contractors or subcontractors. *Piping, structural, modules are often prefabricated.*

Pre-Commissioning: checking and cleaning the Plant without energizing – verifying compliance with engineering and vendors' recommendation. *The mechanical completion is often the conclusion of the Pre-commissioning.*

Purchase Order: contractual document between Purchaser and Supplier for the supply of materials and/or equipment. *To differentiate between contract and PO is a relatively easy exercise – a PO is made from the engineering developed by the supplier; the engineering of an EPSC contract is prepared from a basic design prepared by the Company.*

Reimbursable contract: a contract where the Company compensates the contractor for all expenses incurred.

RFSU: Ready For Start-Up – certificate to be signed.

Site: it is in general the final site where the Plant will be located, but it can also be any worksite where a part of the job is done.

Warranty period: period defined by contract (one to three years) that maintains an obligation by the supplier or the EPC contractor to replace defective material. *Sometimes this word is understood as Guarantee.*

Work: in general all the works to be performed by contractors and suppliers.

Annexes

- Project preparation development, reviews & audits (ref. Chapter 2)
- Tendering procedures, five main phases (ref. Chapter 6)
- Tendering procedures, final step before award (ref. Chapter 6)
- Pre-commissioning & commissioning rule (ref. Chapter 10)
- From construction to start-up (ref. Chapters 6 and 10)
- From construction to production (ref. Chapters 6 and 10)
- Construction examples

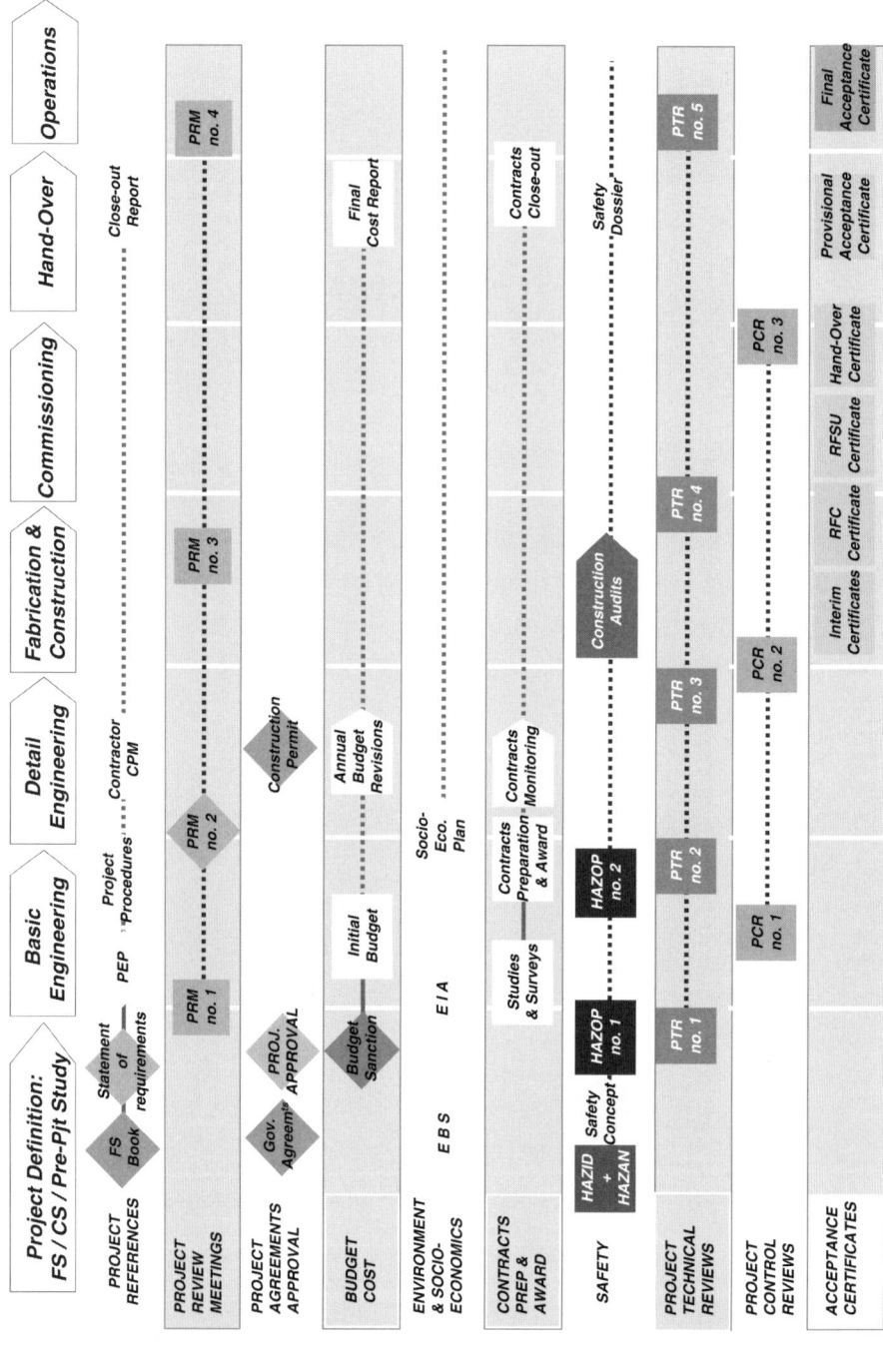

Annexes

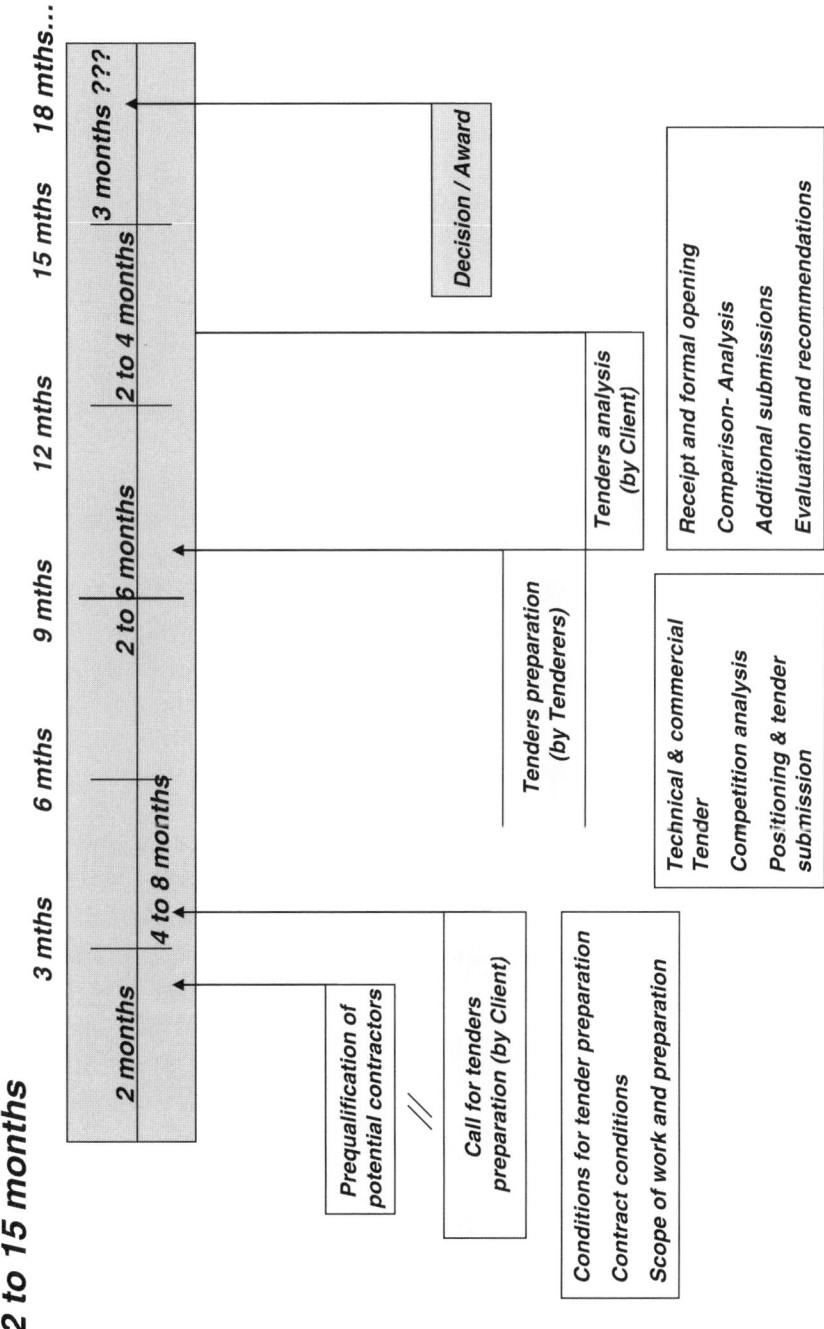

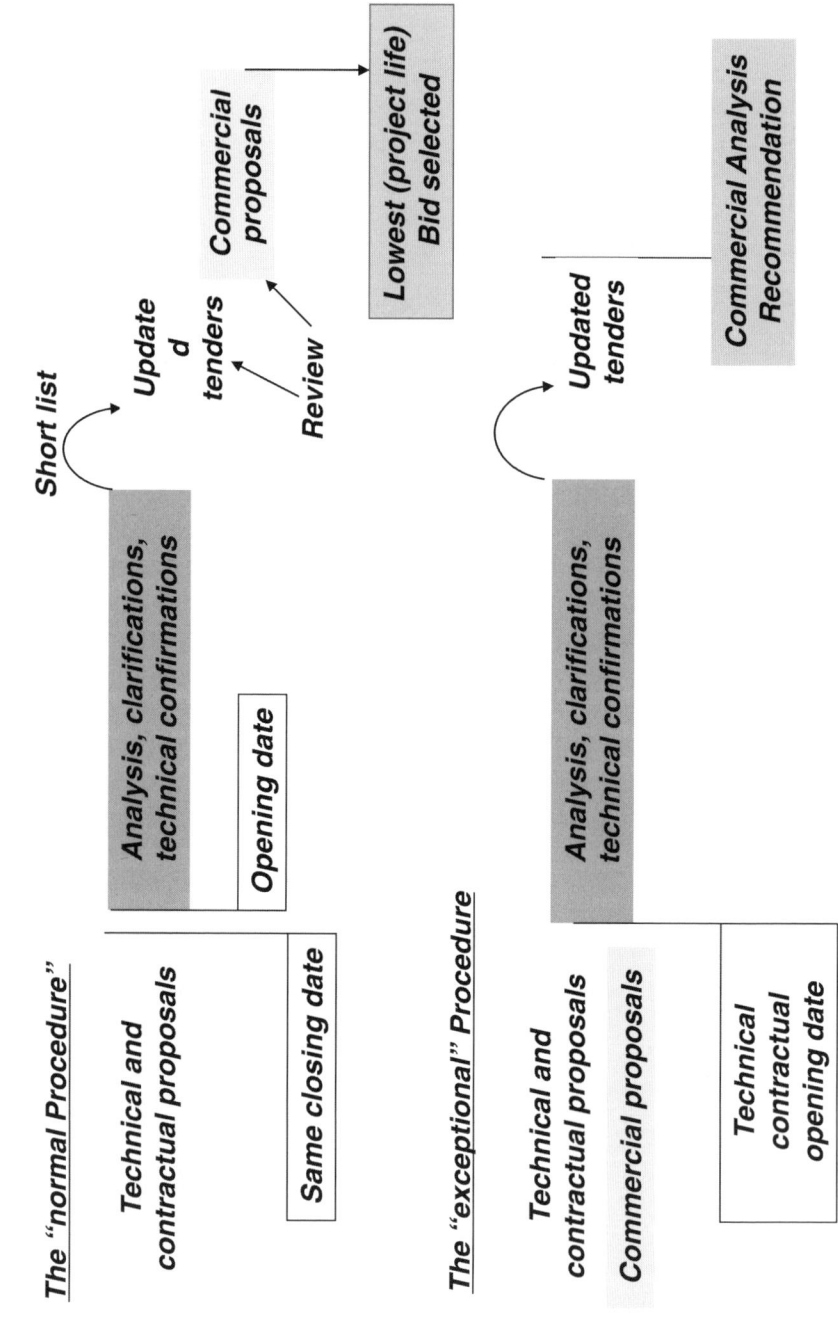

PRE-COMMISSIONING & COMMISSIONING RULE

TRANSFER OF REMAINING WORKSCOPE FROM DISCIPLINES TO SYSTEMS BREAKDOWN

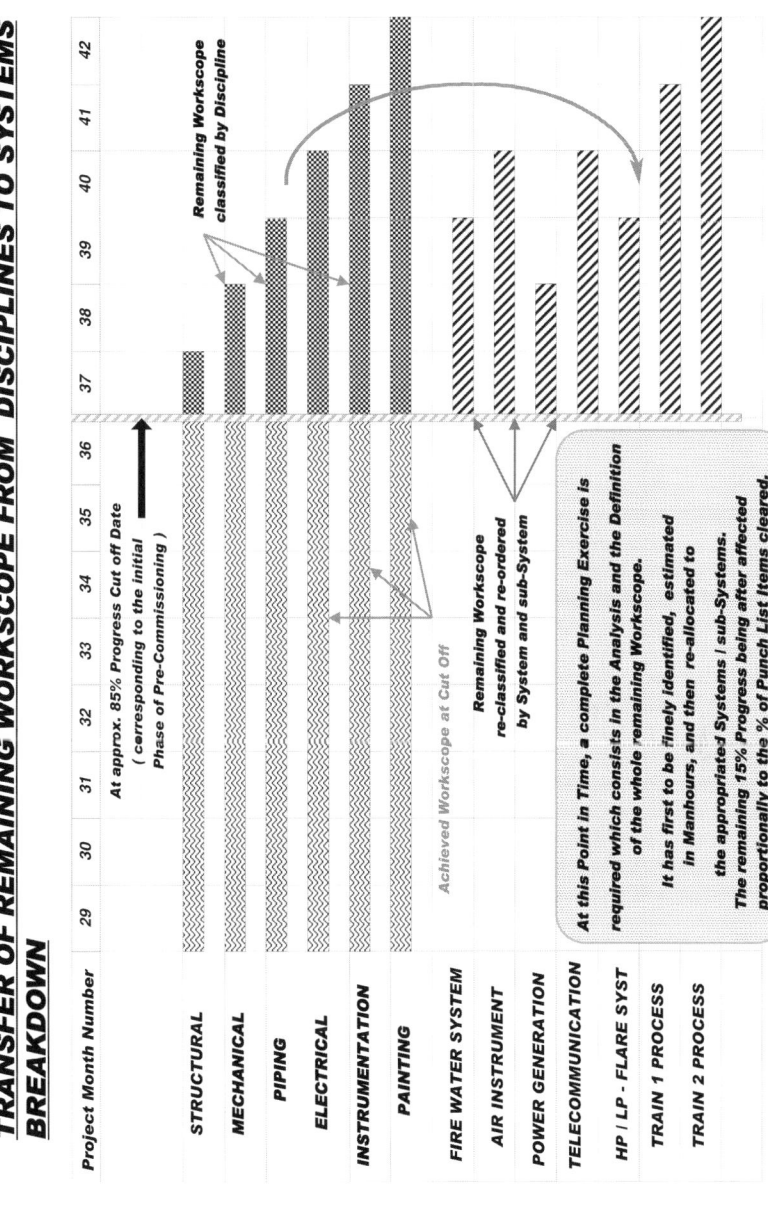

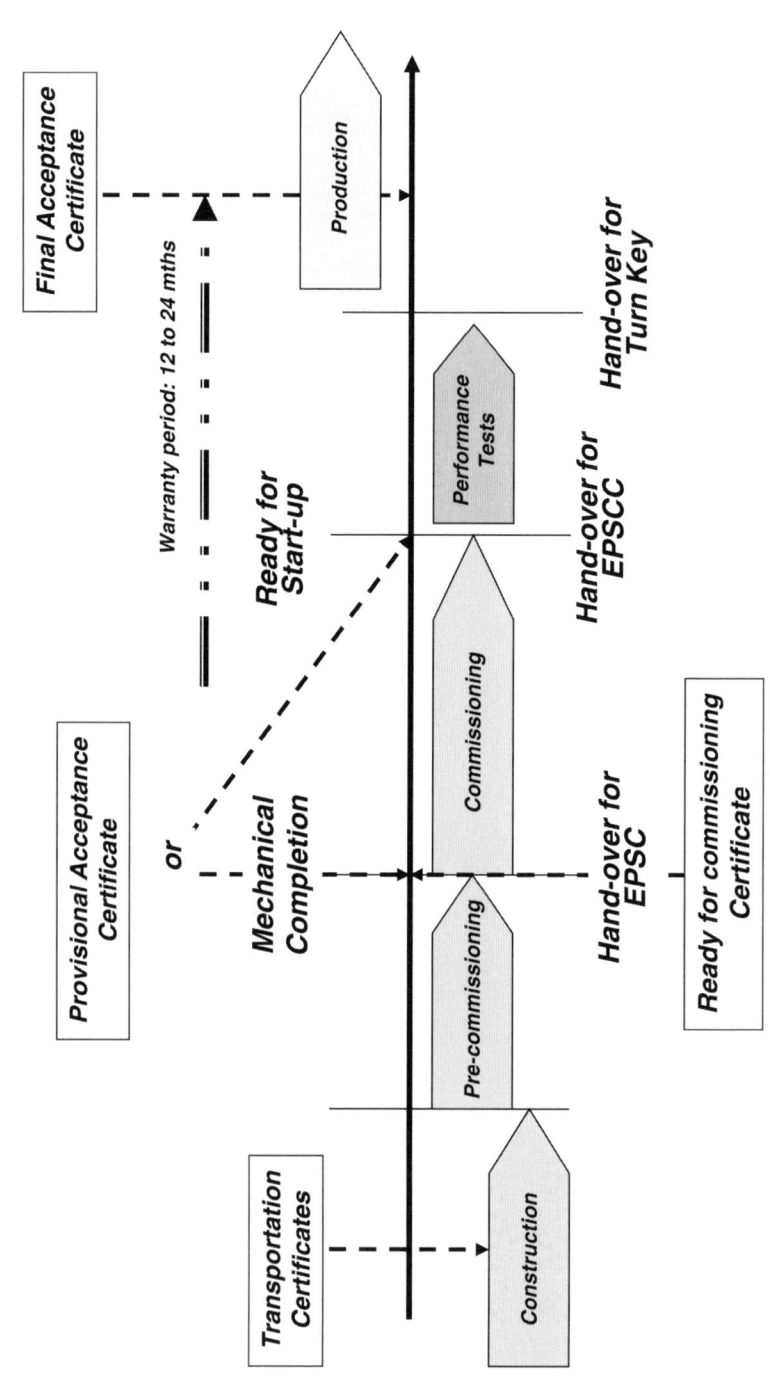

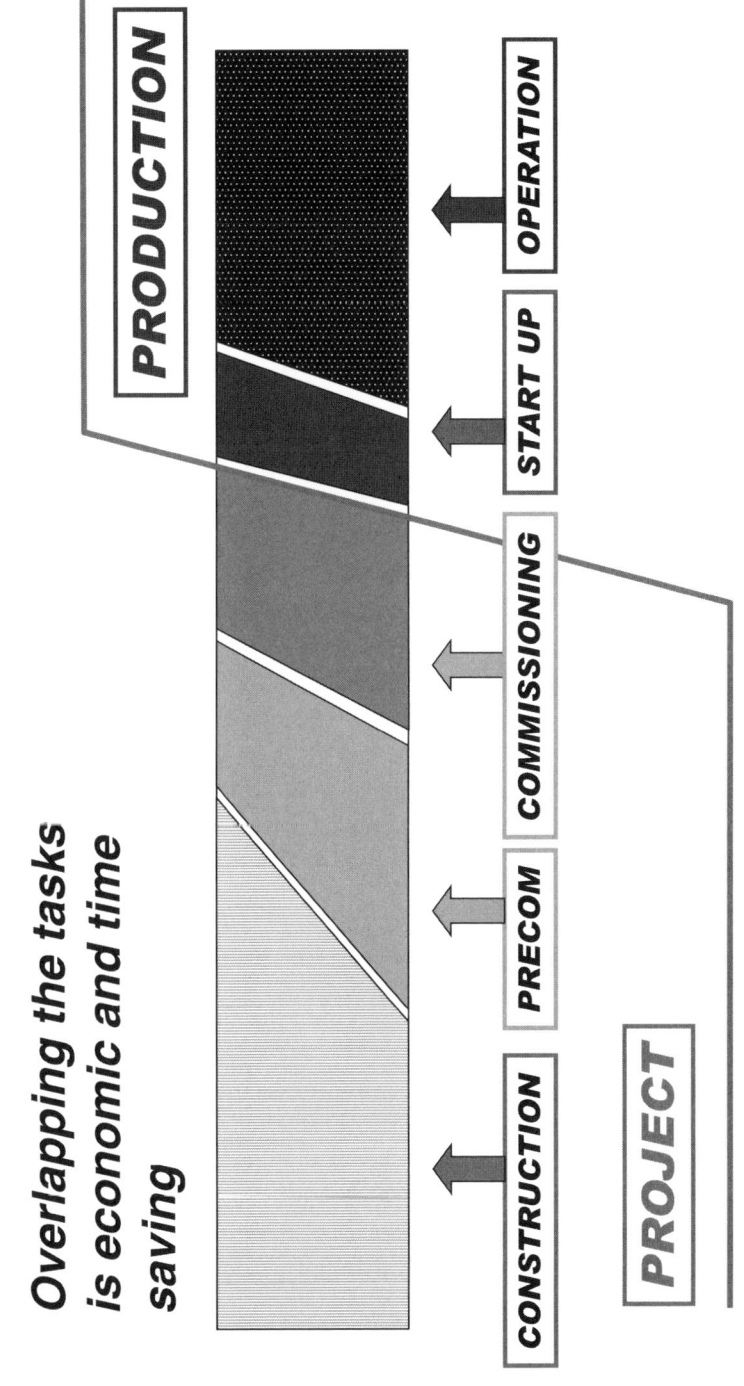

CONSTRUCTION EXAMPLES

Top: QP. (Photo : MD)
Bottom: Jacket and desk construction, Batam. (Photos : MD)

About the Authors

Marc Ducros

Master's degree in physic and chemistry from Paris University, French Petroleum Institute Degree – Refinery (1967).

Successively Assessment Manager in the E & P Division, Service Manager, Project Manager, Project Development Manager, and Gas/Electricity Division Representative in South East Asia for TotalFinaElf, Total-Fina then TOTAL (1976 to 2003).

Secondment to Morocco, UHDE France (1975 to 1976) Scheduling and assessments, Technip France (1969 to 1975).

Teaching and Training 2004 to date:

- Conferences in France: ENSIETA in Brest and ISTIA in Angers (end of 2003).
- Conferences in Thailand: AIT, Management of Large Projects (Feb 2004).
- Short Courses at Chulalongkorn Petroleum College on Gas & LNG (2005 to 2008).
- Short Courses on Gas and LNG in Turkey, in France (IFP).
- Conference on Hydrocarbons challenge for XXI century at Rouen Business School.
- Workshop assistance at ENSPM (2004 to 2008).
- Project Management courses for TPA (Total Professeurs Associés) in Venezuela, Kazakhstan, Iran, Cambodia, Vietnam, China, Russia, Angola.

Gabriel Fernet

Master in Law (Paris Nanterre University).

Sciences Po (Paris University).

Business administration (IAE Paris).

Experience 1973-2003: Lawyer and negotiator with Total. After negotiation of exploration permits (Uranium, Oil & Gas), mainly involved in Oil & Gas development projects, including Alwyn (UK), Idra (Argentina), Yadana Project (Myanmar), Qatargas (Qatar), South Pars (Phase 2 & 3 – Iran), Sincor (Venezuela), etc.

Main Activities:

- Development contract negotiation.
- Development and operation contract and project monitoring (company claims, legal and contractual problem analysis, and follow-up of insurance-related damage and tax issues, etc.)

Teaching and Training:

- A presentation on development contracts (as coordinator of the "Technical Contracts" club in Total's upstream legal department).
- Presentations on insurance, legal issues, etc. at Total.

Project Management courses for TPA (Total Professeurs Associés) in Jordan (for Iraki students), Indonesia, Iran and China.

Imprimé en France en septembre 2013 par EMD S.A.S.
53110 Lassay-les-Châteaux
N° d'imprimeur : 28602 - Dépôt légal : septembre 2010